# Environmental Research and Development

*For Nancy and Kristen*

# Environmental Research and Development

US Industrial Research, the Clean Air Act and Environmental Damage

John T. Scott

*Dartmouth College*

**Edward Elgar**
Cheltenham, UK • Northampton, MA, USA

© John T. Scott, 2003

*Can*

Published by
Edward Elgar Publishing Limited
Glensanda House
Montpellier Parade
Cheltenham
Glos GL50 1UA
UK

Edward Elgar Publishing, Inc.
136 West Street
Suite 202
Northampton
Massachusetts 01060
USA

A catalogue record for this book
is available from the British Library

**Library of Congress Cataloguing in Publication Data**

Scott, John T., 1947 -
     Environmental research and development : US industrial research, the Clean
     Air Act, and environmental damage / John T. Scott.
         p. cm.
     Includes bibliographical references and index.
     1. Air quality – United States. I. Title.

TD883.2.S36 2003
363.739'2'0973–dc21

                                                        2002041388

ISBN 1 84376 167 X

Printed and bound in Great Britain by MPG Books Ltd, Bodmin, Cornwall

# Contents

# List of Figures

# List of Tables

# Preface

As an industrial organization economist with research in the area of R&D and technological change, I began this environmental R&D project as a new departure. I wanted to apply my interests in the industrial organization of R&D and technological change, but do so in an entirely new area for me and in an area of great social importance. I decided to focus on what I have called 'environmental R&D'. The environmental area had not documented in any systematic way, from the perspective of industrial organization, the 'environmental R&D' that industry does. That information interested me for its potential importance to society. Rather than focus on cleaning up the environmental problems caused by existing products and processes, why not do R&D to create new products and processes that provide goods and services without the byproduct of pollution? I wanted to know if industry was working on that approach to improving the environment by avoiding toxic releases with innovative products and processes. With that goal, I designed a survey in 1993 and sent it to all of the significant (by the leading publicly available standard of the time — Business Week's R&D Scoreboard companies) R&D performing companies in the US. I also wanted, after getting industry's response, to wait until the end of the decade before writing the book, so that I could ask if the R&D that industry reported was correlated with a reduction in toxic releases into the environment. Did the R&D have the desired effect? Enough time has passed to answer the question affirmatively in the broad way that my data allow. Further, I have contacted again a substantial group of the initial respondents and reported their update on the environmental R&D being done in industry. I have also presented evidence and theory supporting the use of more aggressive public policies — that I identify and describe — to use the direction of our government to stimulate industry to use its creative powers to improve the environmental performance of industry.

I thank the Nelson A Rockefeller Center for the Social Sciences at Dartmouth College for the financial support of a Reiss Family Senior Faculty Research Grant. I am also grateful for financial support from the Lewis H Haney 1903 Endowment in Economics. I thank Troy J Scott for research assistance on all parts of the project, and Clark HI Khayat for research assistance with the gathering of information from the *Federal Register* about the filings under the National Cooperative Research Act. I am also grateful to the librarians of Feldberg Library and Baker Library of the Dartmouth College Library for guidance with government documents and computerized reference resources and to John F Hawkins of Dartmouth's Computing Services for help with both hardware and software used throughout the project. I also thank William L Baldwin, Albert N Link, FM Scherer, editors Alan Sturmer and Bob Pickens, and three anonymous referees for valuable suggestions that improved the book.

JTS
December, 2002
Hanover, New Hampshire

# 1. A Lesson from History

This book examines the industrial research and development (R&D) response to industry's releases into our environment of hazardous chemicals. In the book, history is juxtaposed with economic theory and statistics; together the history, theory, and statistics teach a lesson.[1] Industry can solve environmental problems rather than create them. Industry can solve environmental problems as it responds over time to external constraints imposed by appropriate public policy. Based on that lesson, this book proposes an aggressive public policy to use the potential within industry to eliminate the problem of hazardous industrial emissions — to eliminate them, rather than simply trying to clean them up after their occurrence.

Politicians in opposing parties, environmental protection groups, and industry debate environmental policy. This book addresses the creative power of industrial R&D to solve the environmental problems when the debate results in governmental guidance and stimulus for such R&D. The focus of the book is on the R&D to change products and processes — a result of *dynamic* competition rather than *static* competition given the existing set of products and processes. Yet the underlying theme of this book — that firms can meet tighter environmental standards by engaging in emissions-related R&D activities — is consistent with a broad range of existing industrial organization and environmental economics literature that favors the use of flexible regulatory approaches (such as tradable permits programs and other market-based programs) to traditional command-and-control approaches. That literature — including Dales (1968), Montgomery (1972), Tietenberg (1985), and Baumol and Oates (1988) — theorizes that the more flexible the regulatory program, the more ingenuity firms will display in adapting and meeting the new requirements. Such ingenuity can be the direct result of firms' R&D activities. Thus, although the environmental literature has typically focused on solutions given the *static* set of products and processes, this book's focus on *dynamic* solutions from new

---

[1] Schumpeter (1954, p. 12) reasoned that 'history, statistics, and 'theory' . . . together make up . . . Economic Analysis.'

*1*

products and processes is grounded in the literature about market-based solutions to environmental problems. This book will develop new data and test the hypothesis that the emissions-reducing R&D investments of US manufacturing firms increase in response to domestic and international competition in the context of emissions regulation.

The history developed in this book focuses on new primary data — the first systematic data of its kind — from the early 1990s and from the early 2000s about the industrial R&D response to toxic industrial emissions. Those emissions have been substantial. In the early 1990s, about 58 percent of the toxic chemicals released into the environment by US industry went into the air, about 22.8 percent were injected underground, about 10.8 percent released onto the land, and 8.6 percent released into surface water. The US Environmental Protection Agency's (EPA's) *Toxics Release Inventory*, reported that US industry released into our environment 3.2 billion pounds of toxic chemicals in 1992 — the year for which this book first estimates the R&D response to emissions of toxic chemicals.[2,3] Reported releases of toxic chemicals into the environment are a decided understatement of the actual total because not all toxic chemicals are covered in the inventory, because not all industries are required to report, and not all firms in reporting industries meet the requirements for threshold size to require reporting. Among those 1992 releases were '197 million pounds of known or suspected cancer-causing chemicals, the report said, along with 166 million pounds of chemicals that damage the earth's protective ozone layer.'[4]

With hindsight, in 1992 US industry — responding to public concern — was in the midst of a dramatic reduction in the emissions of toxic chemicals into the environment. From 1988, when the EPA's *Toxics Release Inventory* began appearing annually, until 1999, total pounds of toxic releases dropped 45.5 percent for the set of core chemicals and for the set of industries that can be traced consistently over the period. Dramatic reductions occurred from 1988 to

---

[2] Note that it is only happenstance that 3.2 billion pounds is also the amount of the releases of the core set of chemicals in 1988 as shown in Table 1.2. Table 1.2 shows releases for a core set of chemicals for which EPA has consistent time series data.

[3] Some readers may find counterintuitive the measurement in pounds of air emissions. Perhaps an analogy to the solid, liquid, and gaseous states of water is useful. Pounds of the solid, ice, or liquid water are readily understood; there are pounds of evaporated water in the air. Of course, water is not a toxic release, but the principle of measuring its amount in pounds while in the gaseous state is the same for any chemical.

[4] These figures are reported by Schmid (1994) from the EPA's annual *Toxics Release Inventory* covering 1992. Schmid points out that critics have complained that the EPA's Inventory has an incomplete coverage of industries and that the numbers of factories that ignore reporting requirements is unknown. One could also add that the measure of total releases by an industry or in a geographic area is in terms of pounds of pollutants, and because the toxicity of pollutants varies and often is unknown for the form of the releases, the numbers of pollutants released will be of interest. For those reasons, I shall use measures of numbers of pollutants across all industries, and the measures are determined by analysis by industry experts rather than being exclusively self-reported by industry. Where appropriate, I shall also use measures of pounds of pollutants for sampled companies.

1995. During that time, 87.3 percent of the 45.5 percent decline occurred. Evidence developed in this book supports the hypothesis that the historic and large reduction in toxic emissions was directly related to emissions-related R&D — the R&D that is surveyed and analysed in this book.

*Table 1.1. EPA's Toxics Release Inventory for 1999: Emissions in Millions of Pounds[a]*

| SIC code | Industry | Total air emis- sions | Surface water dis- charges | Under- ground injec- tion | On-site land releases | Total on-site releases | Off-site releases | Total on- and off-site releases |
|---|---|---|---|---|---|---|---|---|
| 20-39 | Manuf.[b] | 1175.1 | 253.6 | 199.5 | 323.7 | 1951.9 | 374.6 | 2326.5 |
| 10 | Metal mining | 4.5 | 0.45 | 35.1 | 3934.8 | 3974.8 | 2.2 | 3977.0 |
| 12 | Coal mining | 1.8 | 0.24 | 0.14 | 9.6 | 11.8 | 0.00 | 11.8 |
| 491/493 | Electric genera- ting | 841.9 | 4.5 | 0.00 | 258.1 | 1104.6 | 58.0 | 1162.5 |
| 5169 | Chem. whole- salers | 1.3 | 0.0033 | 0.00 | 0.0013 | 1.3 | 0.65 | 2.0 |
| 5171 | Petrol. storage | 4.0 | 0.044 | 0.00 | 0.015 | 4.1 | 0.17 | 4.3 |
| 4953/ 7389 | Waste: treat & dispose | 0.80 | 0.051 | 22.9 | 220.5 | 244.2 | 43.8 | 288.0 |
| TOTAL | | 2029.4 | 258.9 | 257.6 | 4746.7 | 7292.6 | 479.4 | 7772.0 |

[a]*Source*: US EPA (2001, p. E-2). The time period covered by the 1999 data is the calendar year 1999. To avoid putting too fine a point on these rough data, the EPA's data that are given in pounds have been shown in millions and rounded, so the totals may not sum exactly because of rounding error. Also, for categories where less than a million pounds of toxic chemicals were reported, the first couple of significant digits are shown to indicate the presence of some emissions, although of course they tend to disappear in the totals that are rounded to the nearest one-tenth of one million pounds.
[b]The manufacturing industries, Standard Industrial Classification (SIC) industries 20 through 39, were the original reporting industries for EPA's *Toxics Release Inventory*. Only the manufacturing sector reported toxic releases for the inventory in the years 1988 through 1997. Beginning with the reports for 1998, the manufacturing sector was joined by the other industries shown. Since 1998 the data include reports from federal facilities that have reported to EPA for the inventory since 1994.

Table 1.1 shows the EPA's *Toxics Release Inventory* at the end of the 1990s. Then for the core set of chemicals for industries that have reported consistently since 1988, Table 1.2 shows the decline in toxic emissions throughout the 1990s.

Reductions occurred in all of the on-site release categories, with air emissions decreasing by 61 percent, surface water discharges decreasing by 66 percent, underground injection decreasing by 32 percent, and on-site land releases decreasing by 23 percent.[5] This book explains that these reductions came in response to public concerns motivating EPA's maintenance and monitoring of the *Toxics Release Inventory*.

US EPA (2001, p. E-10) emphasizes caveats about the *Toxics Release Inventory* (TRI) data. Not all sources of release are covered; some toxic chemicals and some industry sectors are not covered. Also, small firms not meeting TRI thresholds do not report. Further, the data are estimated by reporting corporations without mandated monitoring, and the estimating techniques can vary across reporting firms and through the years. 'Thus, while the TRI includes 84,068 reports from 22,639 facilities for 1999, the 7.77 billion pounds of on- and off-site releases reported represent only a portion of all toxic chemical releases nationwide.' (US EPA, 2001, p. E-10)

The Clean Air Act Amendments of 1990 are a focus of this history, because Title III of the Act provides a natural experiment for understanding the industrial R&D response to hazardous emissions described in Tables 1.1 and 1.2. In Title III, the US Congress targeted for regulatory attention a collection of chemicals that are at once harmful and useful — both harmful to our environment but extraordinarily useful in industry. Chapter 2 describes these chemicals and associates them with industry and with environmental problems as well.

This book develops new primary data about the industrial R&D response to environmental problems; then, the new data are described and used for hypothesis tests in econometric models. The book, therefore, does not use case studies. Case studies are an important alternative approach to demonstrating that firms' environmental R&D activities can have a direct impact on emissions reduction. However, in the end, one would have only a collection of examples of how firms have innovated in response to environmental regulations. Such examples, of course, help motivate the hypothesis tests in this book's econometric models of the environmental R&D behavior of manufacturing firms, and the examples help illustrate such behavior. Case studies provide motivation and illustrations of the behavior of individual companies, but this book in contrast uses broad samples of firms and studies the range of their behavior using econometric hypothesis testing. Because new primary data are developed, the correspondence with the surveyed companies does generate some quotable comments that suggest motivations and provide illustration. However, this book's focus is on developing a new data set across a large sample of manufacturing firms, using the

---

[5] US EPA (2001, p. E-7) observes that during the period from 1988 through 1999, the number of forms submitted by industry for the *Toxics Release Inventory* did also decline by 5.5 percent.

data to describe industrial environmental R&D, and testing hypotheses about that R&D.

*Table 1.2. Comparisons of EPA's Toxics Release Inventories for Core Chemicals and the Manufacturing Industries in 1988, 1995, and 1999[a]*

| Emissions type | 1988 | 1995 | 1999 | Percent change 1988–99 |
|---|---|---|---|---|
| On-site Air Emissions | 2180.6 | 1204.2 | 858.5 | -60.6 |
| On-site surface water discharges | 41.9 | 17.0 | 14.3 | -66.0 |
| On-site underground injection | 161.9 | 154.7 | 109.3 | -32.5 |
| On-site land releases | 405.9 | 268.3 | 311.9 | -23.1 |
| Total on-site releases | 2790.4 | 1644.3 | 1294.0 | -53.6 |
| Total off-site releases | 422.7 | 293.3 | 455.7 | +7.8[b] |
| Total on-site and off-site releases | 3213.1 | 1937.6 | 1749.7 | -45.5 |

[a]*Source*: US EPA (2001, p. E-9). Totals (in millions of pounds) may not sum exactly because of rounding error.
[b]The increase here is because of large increases in the subcategories for solidification/stabilization and in wastewater treatment (in both cases for metals and metal compounds) and in underground injection and in land treatment and other land disposal.

My survey of US industry reveals the industrial R&D response to public concern about Title III emissions. The survey also allows description of industry's environmental R&D more generally. Chapter 3 describes the survey and the respondents. Chapter 4 describes the R&D response revealed by the survey. The survey shows that 51.7 percent of respondents reported that their R&D on new production processes to lessen toxic emissions was a response to a specific government regulation. Further, 61.5 percent of the respondents stated that their R&D on new products to lessen emissions was a response to a specific government regulation. The percentages suggest that for the survey participants, a non-trivial amount of R&D activity aimed at emissions reduction is spurred on, at least in part, by existing or anticipated environmental regulations.

An economic model of industrial R&D in response to public concern about emissions provides a theoretical basis for several hypotheses. Among these is the hypothesis that the R&D response will be greater, other things being the same, when competitive pressures on industry are greater. Chapter 5 presents that model, and then Chapter 6 tests the hypothesis that competitive pressures increase industrial environmental R&D given the public concerns about emissions.

Cooperation among industrial firms could reduce the competitive pressures that stimulate R&D to reduce emissions, yet the cooperation could lower the costs of the R&D and improve the chances for its success. Chapter 7 examines and interprets the facts about cooperative R&D to reduce hazardous industrial emissions.

Chapter 8 begins by juxtaposing my survey data with the EPA data to show that across industries environmental R&D is correlated with the reduction in toxic releases into the environment. Next, the chapter uses the EPA data and a follow-up survey conducted during 2001 and finds that the pace of environmental R&D and the pace of emissions reduction may have decreased. Certainly the evidence does not suggest an increase in environmental R&D and emissions reduction, despite the remaining large amounts of toxic releases by industry. The 2001 survey was conducted before the events of September 11, 2001, and the aftermath. In the wake of the new priorities created, *reduction* in the discretionary R&D activity of environmental R&D seems likely unless government policy is used to stimulate such R&D.

Finally, the history of the industrial R&D response to society's concerns about industrial damage to the environment suggests that aggressive public policy can use the creative strength of industry to eliminate hazardous industrial emissions. The theory and evidence herein support the use of more stringent regulatory expectations and the use of periodic innovation taxes. Chapter 8 concludes by describing such policies using the lesson from industrial history told in this book.

# 2. Hazardous Industrial Chemicals

## THE INCIDENCE OF TITLE III HAZARDOUS AIR POLLUTANTS

The Clean Air Act Amendments of 1990 (CAAA) listed hazardous air pollutants (emitted by industry) with the intent that they become targets of public scrutiny for future regulation. The CAAA listed the targeted set of hazardous air pollutants in Title III and charged EPA with developing emissions standards to protect the environment and thereby to protect public health. For example, ethylene oxide was listed in Title III in 1990, and in 1994 EPA announced proposed national emissions standards for the chemical.[1]

In the survey to be described in Chapter 3, R&D directors were questioned about whether their company's R&D activity related to Title III chemicals was in response to a particular government regulation. In their marginalia on the questionnaire were comments such as: 'The research is in anticipation of regulation,' and 'We are trying to get out front and avoid future problems.'

As explained in the appendix to this chapter, data from the EPA were used to describe the correspondence between toxic air emissions and industries. Then subsets of these data were created for just the hazardous chemicals — the Title III hazardous air pollutants — that came under increased regulatory scrutiny because of the CAAA.

---

[1] 'The proposed standards would limit emissions of ethylene oxide (EO) from existing and new commercial sterilization and fumigation operations. The proposed national emission standards for hazardous air pollutants . . . implement section 112(d) of The Clean Air Act . . . . The intent of the proposed standards is to protect public health by requiring existing and new major sources and existing area sources to control emissions to the level achievable by the maximum achievable control technology . . . . and by requiring new area sources to control emissions using generally available control technology.' (US General Services Administration, 1994, p. 10,591). 'The proposed standards would reduce nationwide emissions of EO from existing commercial EO sterilization and fumigation facilities by about 93 percent in 1997 compared to the emissions that would result in the absence of the proposed standards. In the absence of a regulation, existing commercial EO sterilization and fumigation operations are projected to emit 1070 Mg (1180 tons) of EO in 1997.' (US General Services Administration, 1994, p. 10,595).

*Figure 2.1. Number of Title III Toxic Air Pollutants by Two-Digit Industries*

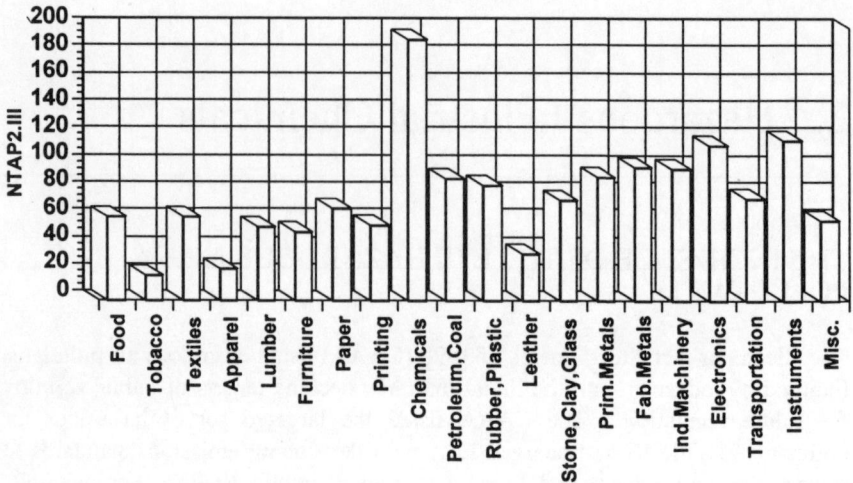

The list of Title III pollutants is provided in US Congress (1990, pp. 2532–5) and is shown in augmented form in the appendix to this chapter. Title III lists 190 chemicals (e.g., benzene) or groups of chemicals (for example, arsenic compounds) as shown in the questionnaire provided in the appendix to Chapter 3. The list in the appendix to Chapter 2 differs from that in Title III's list because Congress's list has been augmented with various specific chemicals in the EPA data base when they belong to a group of chemicals listed in Title III.    For example, arsenic pentoxide is in the data base and is an arsenic compound; and thus, the appendix to this chapter lists arsenic pentoxide *and* a residual category for arsenic compounds not listed separately.

Using the list in this chapter's appendix, there were 205 different Title III toxic air pollutants or families of pollutants emitted by manufacturing industries *and* associated with industries in EPA's data set.   There were 8836 observations of a Title III pollutant associated with a four-digit SIC manufacturing industry.  There were 1408 observations of a Title III pollutant associated with a nonmanufacturing industry.   Thus, about 86 percent of the Title III emissions observations were in manufacturing industry.    Title III emissions — with an 'emission' defined as an observation of a pollutant associated with a four-digit industry — are then about 42 percent (8836/21,004) of manufacturing emissions and about 43 percent (10,244/24,053) of total emissions.  A series of figures can be used to provide a visual representation of the emissions across industries.

*Figure 2.2. Comparison of the Number of Title III Toxic Air Pollutants and the Total Number of Toxic Air Pollutants by Two-Digit Industries*

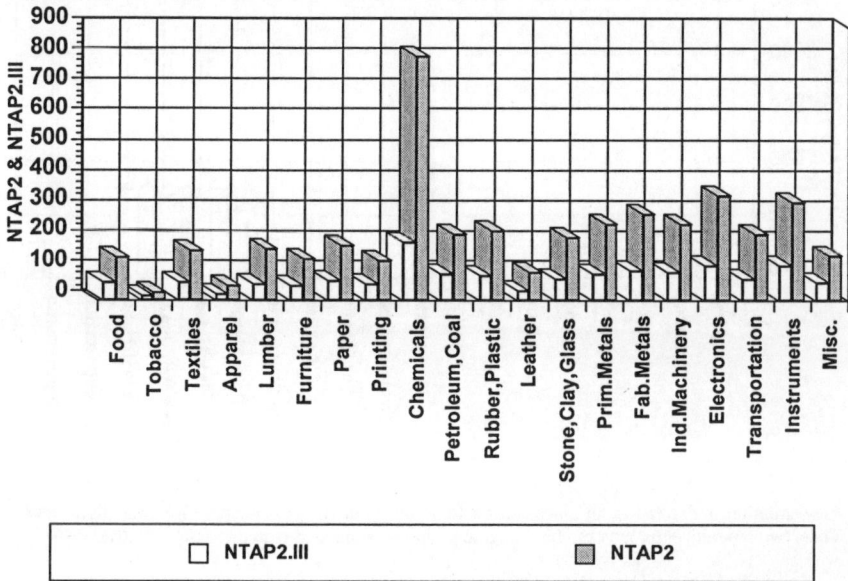

Figures 2.1 through 2.4 describe these data visually. Figure 2.1 shows NTAP2.III, the number of Title III toxic air pollutants associated with each two-digit SIC manufacturing industry. By observing the relatively low number of pollutants associated with the tobacco industry and the relatively high number associated with the electronics industry, one can quickly appreciate that these EPA data show toxic air emissions during the manufacturing process and in the use of the manufactured products in those processes, rather than the toxic emissions during final consumption of consumer goods.

Figure 2.2 compares by two-digit manufacturing industry the number of Title III air toxic pollutants with the total number of pollutants NTAP2. The Title III pollutants are a subset of all pollutants that Congress decided to focus on because it believed existing regulations were inadequate. The comparisons by industry illustrate the relative importance by number of the Title III pollutants. As observed in Chapter 1, numbers of pollutants are of interest given that the information about pounds of pollutants is incomplete and, additionally, the toxicity of the pollutants varies and is often unknown for the various forms of the releases.

Figure 2.3 shows N[TAP,4D]2.III, the number of Title III 'emissions' by two-digit manufacturing industry, where an 'emission' is defined as an instance of a

*Figure 2.3. Number of Title III Toxic Air Emissions\* by Two-Digit Industries*

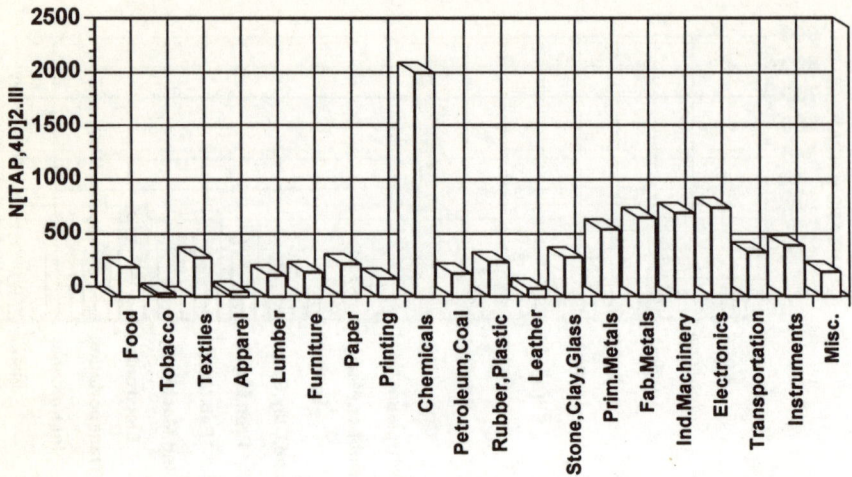

\*An emission is defined as an instance of a toxic air pollutant associated with a four-digit industry. Thus, two separate emissions can be the same pollutant associated with different industries.

toxic air pollutant being associated with a four-digit industry. Thus, a given pollutant can be associated with more than one 'emission' in a two-digit industry.

Figure 2.4 compares Title III 'emissions' with total 'emissions' N[TAP,4D]2. Although for the visual description, the observations have been aggregated to the level of two-digit industries, the estimation and hypothesis testing in subsequent chapters will make use of the underlying information at the level of the more disaggregated four-digit industries.

Each Title III chemical is typically emitted by numerous four-digit manufacturing industries. Table 2.1 provides a frequency distribution showing the number of Title III hazardous air pollutants within a given range for N4DIG, the number of four-digit manufacturing industries that in some way handle, treat, or produce the chemical during the manufacturing process. Hence these industries potentially could release the chemical into the atmosphere; the chemical is probably emitted somewhere in the vertical chain of production associated with the industry. For chemicals that were not associated with a specific four-digit industry but instead were assigned to a more aggregative two- or three-digit industry, the number of industries here counts the two- or three-digit industry as one four-digit industry. Thus, the count here is a lower bound whenever a chemical is associated with more than one of the four-digit industries

*Figure 2.4. Comparison of the Number of Title III Toxic Air Emissions\* and Total Number of Emissions*

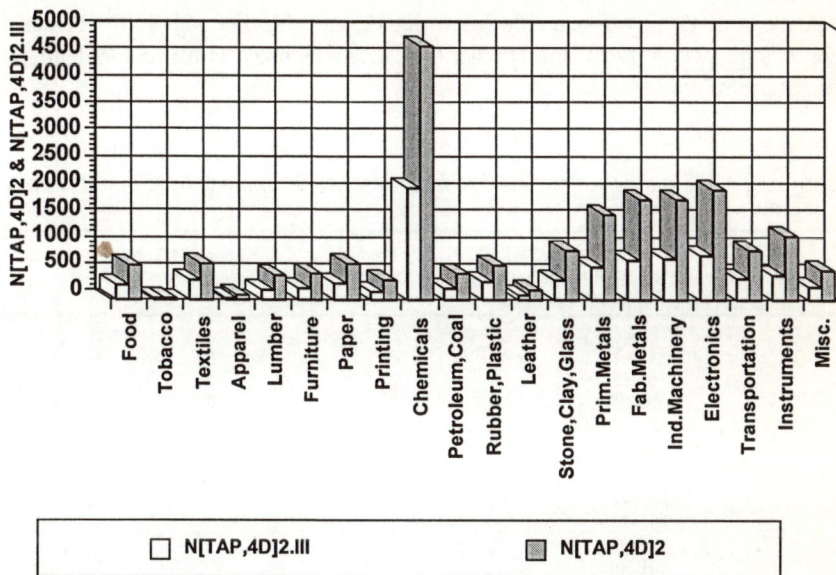

*An emission is defined as an instance of a toxic air pollutant associated with a four-digit industry. Thus, two separate emissions can be the same pollutant associated with different industries.

underlying the more aggregative category.

For example, formaldehyde has one record for Standard Industrial Classification (SIC) industry 23, and, with the observation coded as 2300, that is counted conservatively as one 'four-digit SIC' manufacturing 'emission.' Undoubtedly, however, more than one of the 31 underlying four-digit industries in SIC 23, 'Apparel and Other Textile Products,' uses processes involving formaldehyde.

Some chemicals are associated with the production processes of very few SIC industries. For example, 1,2-Dibromo-3-chloropropane is associated with just one industry. On the other hand, a few chemicals are quite pervasive throughout manufacturing industry.[2]

---

[2] Whenever necessary to pick up a record of a chemical's presence that could be traced only to the more aggregative level, the two-digit categories such as 20, 'Food and Kindred Products,' and three-digit categories such as 202, 'Dairy Products,' were coded as 2000 and 2020 respectively to pick up the record of a toxic chemical associated with one or more of the underlying four-digit industries.

For examples of the pervasive hazardous air pollutants, note that formaldehyde is associated with 224 of the manufacturing industry categories, methylene chloride with 310, methyl ethyl ketone with 341, trichloroethylene with 273, toluene with 427, hydrogen chloride with 313, and chlorine with 250.[3] Toluene, for example, is a solvent among other things; any industry with a tool to degrease might use it.[4]

*Table 2.1. The Incidence of Title III Hazardous Air Pollutants in Manufacturing*

| Frequency Class: A Range for N4DIG, the Number of Four-digit Manufacturing Industries that Emit the Chemical | Frequency: The Number of Title III Chemicals in the Range |
|:---:|:---:|
| 0 – 5 | 47 |
| 6 –15 | 56 |
| 16–25 | 27 |
| 26–50 | 31 |
| 51–75 | 20 |
| 76–100 | 4 |
| 101–150 | 5 |
| 151–200 | 2 |
| 201–300 | 6 |
| 301–400 | 5 |
| 401–450 | 2 |
| The Sum of the Class Totals = Total Number of Cases of a Pollutant Associated with a Particular Industry = 8836 | Total Number of Chemicals = 205 |

The cases where the association of a chemical with an industry are made for the more aggregative two- or three-digit levels are relatively infrequent, but because of those assignments, the frequency distribution of Table 2.1 is just an approximation to the underlying reality about pervasiveness. However, when the information in the related form of NTAP, the number of Title III toxic air pollutants associated with a particular four-digit industry, is used in the statistical

---

[3] Although the formaldehyde example for SIC two-digit industry 23 implies that the industry count for each chemical is a lower bound, for the more pervasive chemicals there may well be an offsetting tendency in the EPA data set to overestimate the count, recording an emission for a four-digit industry and then also recording one for that industry's three- or two-digit aggregative industry even though no other four-digit industries in the aggregative categories are associated with the chemical.

[4] The descriptions of the chemicals make their pervasiveness understandable. For example, toluene is 'a light mobile liquid aromatic hydrocarbon . . . that is produced commercially from light oils from coke-oven gas and coal tar and especially since World War II from petroleum . . . , and that is used chiefly as a solvent, as a raw material for trinitrotoluene, dyes, pharmaceuticals, and other organic compounds, and as a blending agent for gasoline especially for use in aviation . . . .' (*Webster's*, 1971, p. 2405).

work in subsequent chapters, only the exact, unambiguously associated chemicals for a four-digit industry are counted. NTAP is used for subsequent descriptions and hypothesis tests about R&D behavior at the four-digit industry level, and the relatively few cases of chemicals assigned at the two- and three-digit level are not used in the analysis of behavior at the four-digit level.

We can summarize the information about the Title III pollutants and US manufacturing. On average, each of the 205 pollutant categories is associated with 43.10 (8836/205) four-digit industries, and each four-digit industry is associated with 19.25 (8836/459) chemicals. The average value of NTAP for those cases where a chemical is assigned unambiguously to a four-digit manufacturing industry is 16.64.

## MEDICAL EFFECTS OF THE TITLE III CHEMICALS.

The hazards that the Title III chemicals pose to our environment and our health are not fully known, but Congress has formulated public policy given the belief that the hazards are real and substantial. The concern of the medical community is apparent. To illustrate such concern, during the summer of 1993, when the survey described in Chapter 3 was administered, the computerized abstracts of the contents of all medical journals for 1993 and the previous four calendar years were searched systematically for articles that examined the effects of the Title III chemicals.[5]

There are a variety of ways that one can search, each yielding a somewhat different answer. The biggest differences result from different approaches to searching for references to the several cases where Title III specifies a group of chemicals rather than an individual chemical. Table 2.2 provides the frequency distribution showing the number of chemicals within a given range for MED1, MED2, and MED3, alternative measures of the number of medical journal articles studying the chemical. MED1 cast a broader net that was more likely to turn up references to articles for the cases of groups of chemicals (such as 'arsenic compounds') than the narrower search used to compute MED2.

MED1 reflects a search by chemical number or by chemical name. It was a search by chemical number when the chemical had a formal chemical abstract number and was catalogued by number in MEDLINE. Additionally, it was a search by chemical name when there was a number but it was not used in MEDLINE or, in the cases of a group of chemicals where there is no number, by

---

[5] The computerized collection of abstracts (National Library of Medicine, 1993) was available on the Dartmouth College Information Service as the data base 'MEDLINE — 5 years' which is a subset of the National Library of Medicine's MEDLINE file of abstracts of articles from medical journals. The search was conducted in July, 1993.

the basic chemical's name ('arsenic', e.g., when searching for references to arsenic compounds).

MED2 is like MED1 except that it used a 'narrow name search by topic' (for 'arsenic compounds,' for example) for the cases where groups of chemicals are specified in Title III. The two measures MED1 and MED2 have a simple correlation of 0.58. For the 205 Title III chemicals, the correlation of N4DIG, the number of four-digit industries that emit the chemical, and the number of medical journal articles MED1 studying the chemical is 0.082. The correlation of N4DIG with MED2 is 0.25, also positive but quite significant. Because the searches by name rather than by chemical number appear to be likely to pick up many articles that have relatively minor references to a chemical, I have also examined MED3. MED3 simply substitutes a missing observation code for all cases where I had to search by name rather than by number. The sample of 205 is thereby reduced to 165 observations, and for those 165 observations, the simple correlation between N4DIG and MED3 is positive and significant at 0.26. Thus, medical research about the Title III chemicals is positively correlated with their incidence in manufacturing industries. For example, consider two chemicals for which I could search by the chemical abstract number. Formaldehyde with N4DIG = 224 is cited in the abstracts of 1183 medical journal articles during the period examined, while 1,2-Dibromo-3-chloropropane with N4DIG = 1 is abstracted in 37 articles.

A small sampling of the articles enumerated in Table 2.2 can impart the nature of the articles in the context of industrial emissions problems. Heptachlor is a Title III pollutant, and a medical journal article, found in the search of the literature, studies its residue levels in sera from plywood workers. In another example of medical concern about occupational exposure to a Title III pollutant, an article reports a death caused by methyl bromide (Bromomethane), which is used as a fumigant in the food industry. Trichloroethylene is a Title III pollutant used as an industrial solvent. An article examines the neurological problems of metal degreasers who are exposed to the solvent.

Among other examples of medical articles examining Title III pollutants, one studies lung cancer among workers exposed to benzyl chloride, toluene, and benzotrichloride in a factory manufacturing chlorinated toluenes, while another studies lung cancer among workers exposed to epichlorohydrin and chlorine in dye and resin manufacturing. Exposure to epichlorohydrin also appears to be related to heart disease according to another study. Occupational asthma resulting from exposure to maleic anhydride is the focus of study. An article examines acute poisoning by arsine in a ferrous metal foundry. Numerous articles study lung cancer among coke-oven plant workers. Exposure to benzene and mortality from leukemia is also studied for coke-oven and other coal product workers. Beryllium disease among workers exposed to beryllium in the ceramics industry is studied. Lung cancer deaths among workers in man-made mineral fiber production is the focus of another study.

*Table 2.2.  The Medical Literature about Title III Hazardous Air Pollutants*

## MED1

| Frequency Class:  A Range for MED1 | Frequency:  The Number of Title III Chemicals in the Range |
|---|---|
| 0–25 | 107 |
| 26–50 | 21 |
| 51–75 | 13 |
| 76–100 | 14 |
| 101–200 | 19 |
| 201–300 | 9 |
| 301–400 | 3 |
| 401–500 | 6 |
| 501–600 | 2 |
| 601–700 | 1 |
| 701–800 | 1 |
| 801–900 | 0 |
| 901–1000 | 1 |
| 1001 and up | 8 |

| The Sum of the Class Totals = Total Number of Cases of A Pollutant Associated with an Article = 28,514 | Total Number of Chemicals = 205 |
|---|---|

## MED2

| Frequency Class:  A Range for MED2 | Frequency:  The Number of Title III Chemicals in the Range |
|---|---|
| 0–25 | 108 |
| 26–50 | 23 |
| 51–75 | 16 |
| 76–100 | 15 |
| 101–200 | 21 |
| 201–300 | 8 |
| 301–400 | 3 |
| 401–500 | 5 |
| 501–600 | 2 |
| 601–700 | 1 |
| 701–800 | 0 |
| 801–900 | 0 |
| 901–1000 | 1 |
| 1001 and up | 2 |

| The Sum of the Class Totals = Total Number of Cases of A Pollutant Associated with an Article = 17,746 | Total Number of Chemicals = 205 |
|---|---|

TABLE 2.2 CONTINUED

## MED3

| Frequency Class: A Range for MED3 | Frequency: The Number of Title III Chemicals in the Range |
|---|---|
| 0–25 | 83 |
| 26–50 | 19 |
| 51–75 | 12 |
| 76–100 | 13 |
| 101–200 | 17 |
| 201–300 | 8 |
| 301–400 | 3 |
| 401–500 | 4 |
| 501–600 | 2 |
| 601–700 | 1 |
| 701–800 | 0 |
| 801–900 | 0 |
| 901–1000 | 1 |
| 1001 and up | 2 |

| The Sum of the Class Totals = Total Number of Cases of A Pollutant Associated with an Article = 16,141 | Total Number of Chemicals = 165 |
|---|---|

Table 2.2, along with the few examples just described, illustrates the concern about the effects on human health of the Title III hazardous air pollutants. The concern thereby evidenced is explored in subsequent chapters when my survey data are analysed. Clearly, a very large number of articles in the medical journals have examined effects of the Title III chemicals. The industrial settings for the effects on human health are of special interest here because the examples are tied directly to the workers in the industries surveyed. However, the environmental literature has documented the non-human health damages to the environment that result because of human activity — including damages to animals, forests and vegetation, soil, oceans, atmosphere and climate (Friday and Laskey, 1989; Nisbet, 1991).

# 3. A Survey of Industry

## THE QUESTIONNAIRE

To gather information about the industrial R&D response — conditioned by the laws embodying public policies — to the problem of toxic emissions in manufacturing, a questionnaire was used to ask companies about their environmental R&D efforts. The questionnaire, the cover letter that accompanied it, and its appendix listing the Title III chemicals are all provided in the appendix to this chapter. The materials were addressed to the Director of Research & Development for each company, although in many cases the official title of the responding executive was different, typically reflecting the executive's status as the person responsible for the company's programs to reduce environmental hazards. As stated in the survey materials in the appendix, respondents were assured that the individual company's responses would be kept confidential.

The questionnaires were mailed in August 1993 and completed in August and September of 1993. To gain perspective on the changes in industrial environmental R&D during the 1990s, a brief follow-up questionnaire was mailed in July of 2001 to a subset of the original respondents. Namely, it was mailed to those respondents that had reported R&D investments aimed at reducing emissions of air pollutants. The findings and perspectives obtained from the 2001 survey are discussed in Chapter 8; this chapter focuses on the detailed questionnaire about the early 1990s industrial environmental R&D, and in particular on the R&D response to Title III of the Clean Air Act Amendments of 1990.

The questionnaire first asks about the company's research on toxic emissions per se — the generic, background research that would for example develop information about the amount and toxicity of emissions from an industrial process. Then, the questionnaire asks about R&D for the development of new processes that would lessen toxic air emissions, and finally, there are questions about R&D for new products that when used would have fewer toxic air emissions. For all three sets of questions, there are inquiries about emissions problems and R&D in general as well as about efforts related to Title III hazardous air pollutants in particular. The questionnaire asks for information about the various types of R&D as percentages of the company's total R&D or as percentages of one or another of those various types of R&D.

Within each of the three general types of emissions-related R&D, companies are asked to categorize the goals of the R&D more completely. They are also asked to indicate the source or sources of financing and to specify whether the research is done independently or in cooperation with other companies or the government.[1] Companies are asked to describe appropriability conditions, the importance of efficiencies from large-scale efforts, the riskiness, and the costliness of the toxic emissions R&D projects relative to the company's typical projects. They are also asked to indicate whether or not the R&D is a response to a specific government regulation.

## THE SURVEY SAMPLE

The population for the sample was at the time probably the best-known and readily available sample of R&D-intensive US companies — *Business Week*'s R&D Scoreboard sample for 1993. *Business Week* (1993) provides, among other things including sales, the R&D expenses during the most recent fiscal year as of May 18, 1993 for each of the companies. The companies in the sample are those that at the time had sales of $58 million or more and R&D expenses of at least $1 million. The R&D expenses are (*Business Week*, 1993, p. 105): 'Dollars spent on company-sponsored research and development for the most recent fiscal year, as reported to the Securities & Exchange Commission on Form 10-K.' The expenditure figure 'excludes R&D under contract to others, such as US government agencies.' The data reported in *Business Week* were provided by Standard & Poor's Compustat Services.

For each parent company, the four-digit SIC industries in which it and its subsidiaries operated and its primary four-digit SIC industry were determined.[2]

---

[1] Note that the environmental R&D described here is the R&D *performed* by the respondent, not *necessarily* funded by the respondent, although as the survey results described in Chapter 4 show, almost all of the environmental R&D described by the respondents is 'company financed.' That is, the environmental R&D for each respondent is largely financed by the respondent.

[2] The main source for my information about each company's four-digit SIC industries was volume 1 of Standard & Poor's *Register of Corporations, Directors and Executives 1994*. I also used the 1993 version of S&P's *Register*, but for most of the companies in the 1993 *Business Week* R&D Scoreboard the match with sales was exact when the 1994 *Register* was used. The *Register* lists the four-digit SIC industries for each company in numerical order except that the primary industry is listed first (unless the company is designated as a holding company in which case the four-digit code beginning the list designates a holding company).

The set of parent companies was then the basic sample.[3] For a few of those companies, a satisfactory list of four-digit SIC industries could not be developed and those companies were dropped from the sample along with the Scoreboard companies that were not parents. Of the original *Business Week* R&D Scoreboard sample of 891 companies, 846 companies remained for the survey about the industrial R&D response to toxic air emissions. Of those 846 companies, 722 had a manufacturing industry as the primary four-digit SIC industry.

## THE RESPONDENTS

Of the 846 companies surveyed, 150 answered the questionnaire for an overall response rate of 17.7 percent. That response is quite similar to the response rate obtained by Leyden and Link (1992) in their survey of a sample taken from *Business Week*'s R&D Scoreboard a few years earlier. Their study concerned the less sensitive — in terms of the company's concerns about confidentiality — issue of the US government's support of infrastructure technology that is used by firms.

As explained in Chapter 1, the original EPA data about emissions problems in the early 1990s are for manufacturing industries. Therefore, the hypothesis testing in this book uses only the respondents with manufacturing lines of business (LBs) — a line of business or business unit or LB is defined to be a company's operations in a four-digit SIC manufacturing industry. Of the 150 responding firms, 132 firms had manufacturing lines of business. These 132 manufacturing companies operated a total of 717 manufacturing lines of business; thus, the responding manufacturing companies had on average 5.43 manufacturing lines of business. Table 3.1 provides a frequency distribution showing, for the 132-company sample of respondents, the number of manufacturing lines of business in each of the 20 two-digit SIC manufacturing industries. There were no respondents with operations in SIC industry 21, tobacco products; further, no respondents had activity in SIC industry 31, leather products. Leather tanning was the subject of the book and film, *A Civil Action*;

---

[3] For many companies, some subsidiaries and divisions are listed separately (in alphabetical order in volume 1 of the *Register*) from their parent. The third volume of the S&P *Register* for the year provides an index of ultimate parent companies and lists the subsidiaries and divisions of each. Using that listing, I looked up the subsidiaries and divisions of each responding company and gathered up the four-digit SIC industries that belonged to the parent but that were not listed with the parent in the *Register*'s first volume. Whenever there were questions about a particular observation, I cross-checked the *Register*'s information with that provided by OneSource (1993, 1994). For some difficult cases, I also compared the information in Dun & Bradstreet's *Million Dollar Directory* (1993, 1994).

that industry has not done much R&D (for example, see US Federal Trade Commission, 1981, p. 141, p. 146, and p. 278). At the other extreme, there were

*Table 3.1.  The Frequency Distribution of the Lines of Business across Two-Digit SIC Manufacturing Industries for the 132 Responding Companies with Manufacturing Operations*

| Two-Digit SIC Manufacturing Industry | Number of LBs Represented among the 132 Companies |
|---|:---:|
| 20    Food and Kindred Products | 58 |
| 21    Tobacco Products | 0 |
| 22    Textile Mill Products | 8 |
| 23    Apparel and Other Textile Products | 7 |
| 24    Lumber and Wood Products | 12 |
| 25    Furniture and Fixtures | 12 |
| 26    Paper and Allied Products | 31 |
| 27    Printing and Publishing | 3 |
| 28    Chemicals and Allied Products | 104 |
| 29    Petroleum and Coal Products | 8 |
| 30    Rubber and Misc. Plastic Products | 38 |
| 31    Leather and Leather Products | 0 |
| 32    Stone, Clay, and Glass Products | 8 |
| 33    Primary Metal Industries | 28 |
| 34    Fabricated Metal Products | 58 |
| 35    Industrial Machinery and Equipment | 109 |
| 36    Electronic & Other Electrical Equip. | 99 |
| 37    Transportation Equipment | 52 |
| 38    Instruments and Related Products | 71 |
| 39    Miscellaneous Manufacturing | 11 |
| | Total Number of Manufacturing LBs =717 |

roughly 100 lines of business in each of the SIC industries for chemicals, for industrial machinery, and for electronic equipment.

Comparing Table 3.1 with Table 3.2, one sees that the distribution of industries represented in the response largely reflects the distribution of primary industries for the complete 846 company sample that was surveyed. Table 3.2's distribution of primary industries for the complete sample shows the same pattern of the heaviest accumulation of observations among the industries for chemicals, industrial machinery and equipment, and electronics and other electrical equipment, followed by substantial numbers of observations in instruments, and then considerable numbers for food, fabricated metal products, and transportation equipment. The simple correlation between the numbers of LBs reported in Table 3.1 and the numbers of primary LBs reported in Table 3.2 for the 20 two-digit manufacturing industries is positive and highly significant at 0.92.

*Table 3.2. The Frequency Distribution of the Primary Lines of Business across Two-Digit SIC Manufacturing Industries and Nonmanufacturing for the 846 Companies Surveyed*

| SIC Industry | Number of Primary LBs Represented among the 846 Companies |
|---|---|
| 20 Food and Kindred Products | 21 |
| 21 Tobacco Products | 3 |
| 22 Textile Mill Products | 7 |
| 23 Apparel and Other Textile Products | 1 |
| 24 Lumber and Wood Products | 4 |
| 25 Furniture and Fixtures | 5 |
| 26 Paper and Allied Products | 18 |
| 27 Printing and Publishing | 3 |
| 28 Chemicals and Allied Products | 100 |
| 29 Petroleum and Coal Products | 9 |
| 30 Rubber and Misc. Plastic Products | 24 |
| 31 Leather and Leather Products | 2 |
| 32 Stone, Clay, and Glass Products | 7 |
| 33 Primary Metal Industries | 14 |
| 34 Fabricated Metal Products | 33 |
| 35 Industrial Machinery and Equipment | 167 |
| 36 Electronic & Other Electrical Equip. | 144 |
| 37 Transportation Equipment | 48 |
| 38 Instruments and Related Products | 92 |
| 39 Miscellaneous Manufacturing | 20 |
| – Nonmanufacturing | 124 |
| | Total Number of Primary LBs = 846 |

Recalling the distribution of pollutants across industries as described in Chapter 2, the companies surveyed have their primary industries among those with the greatest numbers of toxic air pollutants. NTAP averages 33.96 for the primary industries of the 722 surveyed companies with their primary four-digit industry in manufacturing, roughly twice the simple average for all four-digit manufacturing industries.

The correlation across two-digit industries of the numbers of responding LBs with the numbers of surveyed LBs suggests that there is no particular tendency

*Table 3.3. Primary Industry Effects on Response: OLS for the 846 Companies in the Sample and with DR as the Dependent Variable*

| Dummy Variable for an SIC industry | Coefficient for the Dummy Variable | t-statistic with 825 degrees of freedom | Probability (t)* |
|---|---|---|---|
| D20 Food | 0.14 | 1.6 | 0.055 |
| D21 Tobacco | − 0.19 | − 0.87 | 0.19 |
| D22 Textiles | − 0.051 | − 0.34 | 0.37 |
| D23 Apparel | − 0.19 | − 0.51 | 0.31 |
| D24 Lumber | 0.31 | 1.6 | 0.055 |
| D25 Furniture | 0.21 | 1.2 | 0.12 |
| D26 Paper | 0.029 | 0.30 | 0.38 |
| D27 Printing | − 0.19 | − 0.87 | 0.19 |
| D28 Chemicals | 0.016 | 0.32 | 0.37 |
| D29 Petroleum | 0.14 | 1.1 | 0.14 |
| D30 Rubber & Plastics | − 0.069 | − 0.81 | 0.21 |
| D31 Leather | − 0.19 | − 0.71 | 0.24 |
| D32 Stone, Clay, and Glass | − 0.19 | − 1.3 | 0.097 |
| D33 Primary Metals | 0.24 | 2.2 | 0.014 |
| D34 Fabricated Metals | 0.019 | 0.25 | 0.40 |
| D35 Industrial Machinery | − 0.086 | − 1.9 | 0.029 |
| D36 Electronics | − 0.034 | − 0.73 | 0.23 |
| D37 Transportation | 0.015 | 0.23 | 0.41 |
| D38 Instruments | − 0.020 | − 0.38 | 0.35 |
| D39 Miscellaneous Manufacturing | − 0.044 | − 0.48 | 0.32 |
| Intercept (non-manufacturing) | 0.19 | 5.7 | 0.00 |

$R^2 = 0.0323$; Adjusted $R^2 = 0.0088$

F for the equation as a whole = 1.377 with 20 and 825 degrees of freedom. The probability of a greater F given the null hypothesis that all effects are zero is 0.1245.

*Given the null hypothesis that the coefficient is zero, it is the probability of a larger t-statistic when the estimated coefficient is positive, and it is the probability of a smaller t-statistic when the estimated coefficient is negative.

for responses to come from industries disproportionately to the sampling of the industries. However, a formal analysis shows that the distinction between responding and non-responding companies is to a modest extent because of their primary industries, although the result is not very significant. Ultimately, formal probit analysis of response will be used. However, to begin, a simple, linear probability model provides a useful, intuitive description of the response.

First, by fitting industry effects in a simple ordinary least squares (OLS) model to explain which firms respond, we find that the simplest analysis of variance model is not quite significant at the 10 percent level. DR is a dummy that equals 1.0 if the firm was one of the 150 that responded by answering the questionnaire and is 0.0 otherwise. The intercept picks up the effect for the nonmanufacturing industry observations, and the coefficient on each of the 20 two-digit manufacturing SIC industry dummies represents the difference from the intercept of the two-digit industry's effect on response. Table 3.3 shows the results.

For descriptive purposes, to point up the differences that exist across the industries, consider the one-tailed tests for the industry effects; the probabilities can be doubled to have the statistically complete two-tailed picture of the results. The individual coefficients for the dummy variables for the food and the lumber and wood industries were positive and significant for a one-tailed test at the 5.5 percent level. The primary metals industry also had a significant positive coefficient, significant at the 1.4 percent level. The stone, clay, and glass industry and the industrial machinery and equipment industry were significantly less likely to respond. The dummy variable for the stone, clay, and glass industry had a negative coefficient significantly less than zero at the 9.7 percent level for a one-tailed test, while the coefficient for the dummy variable for the industrial machinery and equipment industry was significantly less than zero at the 2.9 percent level.

A more formal look at the data can be obtained by fitting a probit model of the fixed effects, although the comparison of the probit model's results with the results from the OLS model shows that the OLS results are virtually the same as those from the more formal model. Table 3.4 shows the results from estimating the probit model. With the OLS estimates, one is tempted to add the coefficient for an industry's dummy to the equation's constant term and speak loosely of the result as the probability that a company in that industry will respond. For example, that procedure for the tobacco industry predicts that the probability of response is zero, and for the lumber and wood industry the procedure gives a probability of response of 0.5. Of course, the estimates thus derived were not constrained to behave according to the formal properties of probabilities; moreover, the assumptions of the classical normal linear regression model do not hold, rendering the OLS estimation just an approximation. It is an extraordinarily good approximation here, as we shall now demonstrate by computing the predicted probabilities of response using each model and then comparing the results.

Estimating the probit model allows a rigorous treatment of the probability estimates. With DR, the 0–1 response dummy variable as the dependent variable, the probit model provides maximum likelihood estimates of the coefficients for a

*Table 3.4. Probit Model for DR, the Decision to Respond to the Questionnaire, for the 846 Companies in the Sample: Estimates of the Coefficients for the Probit Index\**

| Industry | Coefficient | Asymptotic t-ratio (z) | Probability > \|z\| |
|---|---|---|---|
| D20 Food | 0.43 | 1.40 | 0.16 |
| D22 Textiles | − 0.20 | − 0.34 | 0.74 |
| D24 Lumber & Wood | 0.86 | 1.35 | 0.18 |
| D25 Furniture | 0.61 | 1.05 | 0.29 |
| D26 Paper | 0.10 | 0.28 | 0.78 |
| D28 Chemicals | 0.058 | 0.31 | 0.76 |
| D29 Petroleum | 0.43 | 0.96 | 0.34 |
| D30 Rubber & Plastics | − 0.29 | − 0.81 | 0.42 |
| D33 Primary Metals | 0.68 | 1.90 | 0.058 |
| D34 Fabricated Metals | 0.066 | 0.24 | 0.81 |
| D35 Industrial Machinery | − 0.37 | − 2.04 | 0.041 |
| D36 Electronics | − 0.13 | − 0.73 | 0.47 |
| D37 Transportation | 0.053 | 0.22 | 0.83 |
| D38 Instruments | − 0.074 | − 0.37 | 0.71 |
| D39 Miscellaneous Manufacturing | − 0.17 | − 0.47 | 0.64 |
| Constant (Non-manufacturing) | − 0.86 | − 6.69 | 0.00 |

Number of observations = 830
Log likelihood = − 381.48
Likelihood Ratio Chi-squared statistic with 15 degrees of freedom = 21.38 (probability of a greater chi-squared = 0.1252)
Pseudo $R^2$ = 0.0273

*\*Notes*:  D21 for tobacco predicts failure perfectly, and the three observations where the sampled company had SIC 21 as its primary industry are therefore not used.  The appropriate coefficient for the dummy would be negative infinity as far as the data reveal, and the probability of response from a company with its primary industry being tobacco is zero.  Also predicting failure perfectly are D23 for apparel (one observation not used), D27 for printing (three observations not used), D31 for leather (two observations not used), and D32 for Stone, Clay and Glass (seven observations not used).  Thus, there are 830 observations used in the probit model rather than 846.

linear combination of the variables explaining response.  In principle, the estimated linear combination — the probit index — can range from minus infinity to plus infinity, covering the domain of the standard normal probability distribution.  The probit index is then translated to the zero to one range by taking the value of the cumulative normal probability distribution at the value for the index.

Thus, for the lumber and wood industry, the sum of the probit coefficient and the constant term gives 0.0. In that case the cumulative normal distribution yields a probability of response of 0.5, precisely what was obtained using the OLS estimates. Table 3.5 compares a company's probability of response given its primary industry for the OLS and probit estimates. The results are almost identical to two decimal places.

*Table 3.5. The Probability that a Company Will Answer the Questionnaire Given Its Primary SIC Industry*

| Industry | Probability Using OLS Estimates | Probability Using Probit Estimates |
|---|---|---|
| Food | 0.33 | 0.33 |
| Tobacco | 0.00 | 0.00 |
| Textiles | 0.14 | 0.14 |
| Apparel | 0.00 | 0.00 |
| Lumber and Wood | 0.50 | 0.50 |
| Furniture | 0.40 | 0.40 |
| Paper | 0.22 | 0.22 |
| Printing and Publishing | 0.00 | 0.00 |
| Chemicals | 0.21 | 0.21 |
| Petroleum | 0.33 | 0.33 |
| Rubber & Plastics | 0.12 | 0.13 |
| Leather | 0.00 | 0.00 |
| Stone, Clay, and Glass | 0.00 | 0.00 |
| Primary Metals | 0.43 | 0.43 |
| Fabricated Metals | 0.21 | 0.21 |
| Industrial Machinery | 0.10 | 0.11 |
| Electronics | 0.16 | 0.16 |
| Transportation | 0.20 | 0.21 |
| Instruments | 0.17 | 0.18 |
| Misc. Manufacturing | 0.15 | 0.15 |
| Nonmanufacturing | 0.19 | 0.19 |

In order to pursue further the question of whether there is a systematic bias in the sample with regard to the variables of interest, we can try to probe further into why some companies responded and many did not. One might expect that those firms in especially polluting industries, or those with an especially sensitive environmental record, would have been reluctant to respond. The first possibility was tested using the US EPA (1991) data about toxic emissions of Title III chemicals in combination with the information from the Standard & Poor's *Register* (1993, 1994) and OneSource (1993, 1994) about each company's primary industry. The second was tested using the data gathered by the Investor Responsibility Research Center (1993).

Subsidiary hypotheses about the type of firm — its industries and whether or not it was an especially R&D intensive or an especially large firm — were tested using the data gathered from *Business Week* (1993) and from the S&P *Register* (1993, 1994). Various possibilities come to mind. On the one hand, a large firm or a firm with a large R&D operation may be more likely to have a part of its bureaucracy that would handle the questionnaire. On the other hand, the bureaucracy itself might inhibit a response because of a more developed chain of command through which the response must be cleared.

The possibilities that firm size and R&D intensity affect the likelihood that a firm responded to the questionnaire are not supported at all in the complete 846 company sample. Company size measured by sales (and the nonlinear in the variables version of the response model using sales and sales squared), company R&D expenditures, and company R&D intensity measured as R&D divided by sales are not at all significant. First, they have no impact on DR when entered without the industry effects — no matter whether they are entered singly or in various groups. All of those models, OLS or probit, show $R^2$ or pseudo-$R^2$ essentially zero with adjusted $R^2$ being negative. Further, none of the variables have a significant effect when entered into the fixed effects models of Table 3.3 and Table 3.4.

To test the possibility that a marked reluctance to respond characterizes companies in industries where toxic air emissions in manufacturing pose particularly acute problems, we can turn to the sample of 722 firms for which the primary four-digit industry is in manufacturing. As it turns out, such reluctance does appear to be present to a modest extent. As discussed earlier, because information about pounds of pollutants and the toxicity of pollutants is incomplete, we shall consider numbers of pollutants and the medical community's concern about the pollutants. NTAP, the number of Title III toxic air pollutants traced to a four-digit manufacturing industry, for the company's primary industry, and also MED2PR, the average of MED2 for all of the Title III chemicals associated with the company's primary four-digit industry, are both significant determinants of DR, the decision to respond, when entered with the two-digit industry effects and whether the estimation uses an OLS or a probit model. NTAP controls for the number of chemicals, while MED2PR controls for the average level of concern for those chemicals associated with production in the four-digit manufacturing industry. MED2 was used rather than MED1, which as discussed earlier appears to pick up extraneous articles, or MED3, which as discussed earlier has several missing observations and which appears to perform similarly to MED2.

Table 3.6 shows the OLS results; while the probit model's results are shown in Table 3.7. As shown in Tables 3.6 and 3.7, NTAP which reflects the pollution problem and MED2PR which reflects the medical community's concern about the problem are both associated with a lower probability of response.[4] Just as for the complete 846 firm sample, the other variables — size in terms of sales or R&D and R&D intensity — are not significant in any specification and their presence does not change the qualitative findings about NTAP and MED2PR.

In the 722 company sample, the average value for NTAP is 33.96 and its estimated standard deviation is 27.90. So, using the OLS results, as we move from a company with NTAP for its primary industry being one standard deviation below the average to one with NTAP a standard deviation above the average, the probability of responding changes by $- 0.0012*27.90*2$ or $- 0.067$. The average value of MED2PR for the 722 companies is 174.0 with an estimated standard deviation of 51.22. Thus, a company with MED2PR one standard deviation above the average will have a probability of responding that is 0.058 less than the probability for a company with MED2PR one standard deviation below the average.

Using the probit estimates in Table 3.7, consider the probability that a company in the food industry would respond to the survey. The probit index is equal to $0.17 - 0.0048*NTAP - 0.0024*MED2PR$. Using the means and standard deviations for NTAP and MED2PR in the 722 observation sample, the probit index at the averages for the explanatory variables equals $0.17 - (0.0048)*(33.96) - (0.0024)*(174.0) = - 0.410608$, which corresponds to a probability of response of 0.341. If NTAP is a standard deviation above its mean, then the probit index for a company in the food industry with the average value for MED2PR equals $0.17 - (0.0048)*(33.96 + 27.90) - (0.0024)*(174.0) = - 0.544528$, which corresponds to a probability of 0.293. If instead, NTAP is held at its mean and MED2PR is increased to a standard deviation above its mean, then the probit index becomes $0.17 - (0.0048)*(33.96) - (0.0024)*(174.0 + 51.22) = - 0.533536$, corresponding to a probability of 0.297. In sum, the effect of a standard deviation increase in NTAP is to reduce the probability of response by 0.041, and the effect of the increase in MED2PR by a standard deviation is to

---

[4] Tables 3.6 and 3.7 include the two-digit industry effects; they clearly belong in the models, although they are only modestly significant, they are significant as a group. However, the effects of NTAP and MED2PR on the probability of response are also present, although not as significant, when they are entered alone, without controlling for the broad industry effects. For OLS of DR on NTAP and MED2PR, the t-ratio for NTAP is $-1.07$ (the probability of a lower t against the null hypothesis is 0.14) and the t-ratio for MED2PR is $- 1.82$ (the probability of a lower t against the null hypothesis is 0.035). The F for the regression as a whole is 1.70 with 2 and 719 degrees of freedom. Given the null hypothesis, the probability of a greater F is 0.18. For the probit model without the industry effects, the asymptotic t-ratios for NTAP and MED2PR are $- 1.06$ and $- 1.82$ respectively and the likelihood ratio test for the model is 3.45 with 2 degrees of freedom. The probability of a greater likelihood ratio given the null hypothesis is 0.18.

*Table 3.6. The OLS Effects of Toxic Chemicals in the Primary Four-Digit Industry and Medical Concerns on the Response DR for the 722 Companies in the Sample with the Primary Industry in Manufacturing*

| Variable | Estimated Coefficient | t-statistic with 700 degrees of freedom | Probability (t)* |
|---|---|---|---|
| NTAP | − 0.0012 | − 1.7 | 0.045 |
| MED2PR | − 0.00057 | − 1.7 | 0.045 |
| D21 Tobacco | − 0.29 | − 1.2 | 0.12 |
| D22 Textiles | − 0.19 | − 1.2 | 0.12 |
| D23 Apparel | − 0.39 | − 1.0 | 0.16 |
| D24 Lumber | 0.19 | 0.91 | 0.18 |
| D25 Furniture | 0.033 | 0.18 | 0.43 |
| D26 Paper | − 0.097 | − 0.80 | 0.21 |
| D27 Printing | − 0.34 | − 1.5 | 0.067 |
| D28 Chemicals | − 0.10 | − 0.93 | 0.18 |
| D29 Petroleum | 0.031 | 0.20 | 0.42 |
| D30 Rubber & Plastics | − 0.23 | − 2.0 | 0.023 |
| D31 Leather | − 0.40 | - 1.4 | 0.081 |
| D32 Stone, Clay, and Glass | − 0.34 | − 2.1 | 0.018 |
| D33 Primary Metals | 0.090 | 0.68 | 0.25 |
| D34 Fabricated Metals | − 0.13 | − 1.2 | 0.12 |
| D35 Industrial Machinery | − 0.22 | − 2.5 | 0.0063 |
| D36 Electronics | − 0.18 | − 2.0 | 0.023 |
| D37 Transportation | − 0.14 | − 1.3 | 0.097 |
| D38 Instruments | − 0.17 | − 1.8 | 0.036 |
| D39 Miscellaneous Manufacturing | − 0.20 | − 1.7 | 0.045 |
| Intercept (Food) | 0.47 | 4.1 | 0.000023 |

$R^2 = 0.0444$;  Adjusted $R^2 = 0.0157$
F for the equation as a whole = 1.548 with 21 and 700 degrees of freedom.  The probability of a greater F given the null hypothesis that all effects are zero is 0.056.

*Given the null hypothesis that the coefficient is zero, it is the probability of a larger t-statistic when the estimated coefficient is positive, and it is the probability of a smaller t-statistic when the estimated coefficient is negative.

reduce the probability by 0.044.

In this chapter, the effects for the OLS 'linear probability' model are juxtaposed with the more refined probit estimates to ensure that the results are presented in an intuitive way.  The foregoing probit predictions of effects are fairly close to the effects predicted by the linear probability model, but they are appropriately sensitive to the starting point for observation for which the change in the variable is observed. For that reason, having in this chapter used the OLS and probit comparisons to illustrate their differences and enhance our intuition, we shall refer only to the probit results in subsequent chapters' descriptions of the predicted effects.

As a final observation about the probability of response, consider a company in the food industry for which both NTAP and MED2PR exceed their mean values by a standard deviation. The probit index would be equal to $0.17 - (0.0048)*(33.96 + 27.90) - (0.0024)*(174.0 + 51.22) = - 0.667456$, corresponding to a probability of 0.252. For such an observation, the probability of responding to the survey would fall below the probability at the sample's average values for NTAP and MED2PR (a 34.1 percent chance) to a 25.2 percent chance.

To test the possibility that companies with an especially sensitive environmental record were reluctant to respond, we turn now to the data gathered by the Investor Responsibility Research Center (1993). IRRC provides information about the environmental performance of companies in the Standard & Poor's 500 index as of December 1990. IRRC (1993) is quite forthright and thorough in listing limitations of the data about environmental performance. For example, (IRRC, 1993, pp. 6–7) the toxic chemical transfers and releases data are 'largely self-reported by the companies,' under the Emergency Planning and Community Right-to-Know Act of 1986, and 'companies have used different methods for estimating emissions.' Earlier estimates have been revised by some companies but not others, and there is 'some ambiguity' about exactly what is reportable. Further, not all facilities are required to report toxic chemical emissions. The different toxicities of the chemical emissions are not accounted for, in part because 'there is little scientific agreement on toxicity for many chemicals.'

Note that the foregoing problems with emissions data are good reasons to focus on experts' counts of chemicals associated with manufacturing industries rather than on emissions data. However, the IRRC's emissions data will surely allow a good check of the hypothesis that an especially sensitive environmental record affects a company's probability of responding to the questionnaire.

For our purposes here, a key piece of information gathered by IRRC is used for each of the 185 IRRC companies that are in the R&D Scoreboard data set and that have their primary industry in manufacturing. The information used is the variable EEI. EEI is IRRC's 'emissions efficiency index' which shows, for the three years 1988–1990, 'the ratio of reported toxic chemical emissions in pounds to the company's domestic revenues (expressed in thousands of dollars). A high ratio may indicate that a company is operating in an industry with relatively high

*Table 3.7. The Probit Model of Effects of Toxic Chemicals in the Primary Four-Digit Industry and Medical Concerns on the Response DR for the 722 Companies in the Sample with the Primary Industry in Manufacturing[a]*

| Variable | Coefficient | Asymptotic t-ratio (z) | Probability > \|z\| |
|---|---|---|---|
| NTAP | – 0.0048 | – 1.66 | 0.097 |
| MED2PR | – 0.0024 | – 1.77 | 0.076 |
| D22 Textiles | – 0.65 | – 0.99 | 0.32 |
| D24 Lumber & Wood | 0.51 | 0.74 | 0.46 |
| D25 Furniture | 0.033 | 0.05 | 0.96 |
| D26 Paper | – 0.30 | – 0.68 | 0.50 |
| D28 Chemicals | – 0.31 | – 0.79 | 0.43 |
| D29 Petroleum | 0.11 | 0.20 | 0.84 |
| D30 Rubber & Plastics | – 0.81 | – 1.82 | 0.069 |
| D33 Primary Metals | 0.22 | 0.48 | 0.629 |
| D34 Fabricated Metals | – 0.42 | – 1.10 | 0.27 |
| D35 Industrial Machinery | – 0.82 | – 2.56 | 0.011 |
| D36 Electronics | – 0.61 | – 1.87 | 0.062 |
| D37 Transportation | – 0.45 | – 1.23 | 0.219 |
| D38 Instruments | – 0.56 | – 1.69 | 0.090 |
| D39 Miscellaneous Manufacturing | – 0.71 | – 1.56 | 0.118 |
| Constant (Food)[b] | 0.17 | 0.39 | 0.70 |

Number of observations = 706; Log likelihood = – 318.05; Likelihood Ratio Chi-squared statistic with 16 degrees of freedom = 26.23 (probability of a greater chi-squared = 0.0509); Pseudo $R^2$ = 0.0396.

[a]D21 for tobacco predicts failure perfectly, and the three observations where the sampled company had SIC 21 as its primary industry are therefore not used. The appropriate coefficient for the dummy would be negative infinity as far as the data reveal, and the probability of response from a company with its primary industry being tobacco is zero. Also predicting failure perfectly are D23 for apparel (one observation not used), D27 for printing (three observations not used), D31 for leather (two observations not used), and D32 for Stone, Clay and Glass (seven observations not used). Thus, there are 706 observations used in the probit model rather than 722.

[b]Since the intercept (Food industry) term is close to 0.5 in the OLS estimation of Table 3.6, then following the discussion earlier about the meaning of the probit coefficients, we expect that, the large t-ratio on the OLS intercept not withstanding, the corresponding constant term in the probit model will not differ significantly from zero. Hence it would not differ significantly from the value of the probit index that corresponds to 0.5 for the cumulative normal density function.

pollutant emissions' (IRRC, 1993, p. 19). IRRC (1993, p. 6) observes, 'While there is no certainty that a pattern of increasing emissions per dollar of revenue produced will result in greater financial risks, proliferating environmental regulations and litigation strongly suggest that companies able to generate revenue with lower levels of regulated pollutant emissions will tend to have fewer future environmental liabilities.' The figure for toxic chemicals transfers and releases in pounds 'is the sum of the reported emissions of selected toxic chemicals from domestic manufacturing facilities owned by the company and its subsidiaries' (IRRC, 1993, p. 18).[5]

EEI should provide, then, a rough inverse measure of a company's environmental performance. Higher values of EEI indicate, subject to all of the caveats noted by IRRC, poorer environmental performance; and thus, with a higher EEI a company might be more reluctant to respond because it is more concerned about potential liabilities. The information requested in the questionnaire is then more sensitive for a company with higher EEI. In fact, Tables 3.8 and 3.9 show that a higher EEI for a company is associated with a lower probability of answering the questionnaire.

Note that EEI is an alternative to NTAP, with EEI providing a company-specific measure of the pollution problem associated with the company's operations. Also, in this much smaller sub-sample for which the IRRC data were available, there are no longer any companies with primary operations in several of the two-digit industries (in particular, SIC industries 22, 23, 25, 27, and 31). The remaining dummy variables that could be estimated were no longer significant as a group. Moving from the OLS model with EEI and MED2PR alone to the model with the remaining dummy variables added as well, the F-ratio for the additional 14 effects is 1.275 with 14 and 168 degrees of freedom. A bigger F has a probability of 0.23 against the null hypothesis. For the 14 two-digit effects estimated alone, the model's F-ratio is 1.363 with 14 and 170 degrees of freedom. The probability of a higher F given the null hypothesis is 0.18.

For the probit model with only the industry effects estimated, of the 15 industries represented in the 185 observation sample, observations from industries 21 (with two observations), 24 (with two observations) and 32 (with two observations) are dropped because they predict non-response perfectly. Then, for the probit model of the industry effects with the remaining 179 observations representing 12 industries, the likelihood ratio chi-squared statistic

---

[5] Note that 'IRRC presents the total amount of reported chemical releases and transfers as one combined figure on the profile. Releases are emitted directly into air, water, land and underground injection wells. Transfers to off-site locations, primarily publicly owned treatment works and private hazardous waste treatment and disposal facilities, are usually treated before being released into the environment. Information on the total amount of chemical releases and transfers is intended to indicate the potential for environmental damage and liabilities resulting from continued emission of pollutants into the environment as well as the potential for increased costs associated with off-site treatment and disposal of hazardous wastes' (IRRC, 1993, p. 6).

is 14.00 with 11 degrees of freedom; the probability against the null hypothesis of a higher chi-squared statistic is 0.23.

*Table 3.8. The Effects of a Company's Environmental Record on Its Decision to Respond DR, for the 185 Company Sample with IRRC's Environmental Efficiency Index, without Industry Effects*

OLS Results

| Variable | Estimated Coefficient | t-statistic with 182 degrees of freedom | Probability (t)* |
|---|---|---|---|
| EEI | − 0.0064 | − 1.6 | 0.056 |
| MED2PR | − 0.00081 | − 1.6 | 0.056 |
| Intercept | 0.33 | 3.5 | 0.00029 |

$R^2 = 0.0221$; Adjusted $R^2 = 0.0114$; F for the equation as a whole = 2.058 with 2 and 182 degrees of freedom. The probability of a greater F given the null hypothesis that all effects are zero is 0.131.

Probit Results

| Variable | Estimated Coefficient | Asymptotic t-ratio (z) | Probability > |z| |
|---|---|---|---|
| EEI | − 0.075 | − 1.56 | 0.12 |
| MED2PR | − 0.0038 | − 1.71 | 0.088 |
| Constant | − 0.15 | − 0.36 | 0.72 |

Number of observations = 185; Log likelihood = − 82.05; Likelihood Ratio Chi-squared statistic with 2 degrees of freedom = 6.30 (probability of a greater chi-squared = 0.043); Pseudo $R^2 = 0.037$.

*Given the null hypothesis that the coefficient is zero, it is the probability of a larger t-statistic when the estimated coefficient is positive, and it is the probability of a smaller t-statistic when the estimated coefficient is negative.

EEI is nonetheless significant with the industry effects estimated, but Table 3.8 presents the results without them since they are insignificant taken as a group. For comparison, with the effects estimated, the coefficient for EEI in the OLS estimation is − 0.00693 with a t-ratio of − 1.59 with 168 degrees of freedom, which is quite similar to the result without the industry effects — a coefficient of − 0.00643 with a t-ratio of − 1.57 with 182 degrees of freedom. For the probit

model with the industry effects estimated as well as coefficients for EEI and
MED2PR, six observations representing three industries where predictions of

*Table 3.9. Probit Models of the Effects of a Company's Environmental Record
on Its Decision to Respond DR, for the 185 Company Sample with IRRC's
Environmental Efficiency Index, with Discernible Industry Effects*

The Full Sample

| Variable | Estimated Coefficient | Asymptotic t-ratio (z) | Probability > \|z\| |
|---|---|---|---|
| EEI | – 0.123 | – 2.06 | 0.039 |
| MED2PR | – 0.00393 | –1.71 | 0.088 |
| D33 (Primary Metals) | 1.834 | 2.14 | 0.033 |
| D35 (Industrial Machinery) | – 0.597 | –1.78 | 0.074 |
| D38 (Instruments) | – 0.455 | – 1.10 | 0.273 |
| Constant | 0.0384 | 0.09 | 0.927 |

Number of observations = 185; Log likelihood = – 77.05; Likelihood Ratio Chi-squared
statistic with 5 degrees of freedom = 16.31 (probability of a greater chi-squared = 0.0060);
Pseudo $R^2$ = 0.0957.

The Sample without the Perfect Prediction Cases

| Variable | Estimated Coefficient | Asymptotic t-ratio (z) | Probability > \|z\| |
|---|---|---|---|
| EEI | – 0.120 | – 2.03 | 0.043 |
| MED2PR | – 0.00317 | – 1.33 | 0.184 |
| D33 (Primary Metals) | 1.771 | 2.07 | 0.039 |
| D35 (Industrial Machinery) | – 0.636 | – 1.89 | 0.058 |
| D38 (Instruments) | – 0.485 | – 1.17 | 0.242 |
| Constant | – 0.0645 | – 0.15 | 0.881 |

Number of observations = 179 since this sample's two observations in each of the SIC
industries 21 (tobacco), 24 (lumber and wood), and 32 (stone, clay, and glass) were not
used because the categorical variable for each of those industries predicts non-response
perfectly; Log likelihood = – 76.23; Likelihood Ratio Chi-squared statistic with 5 degrees
of freedom = 15.64 (probability of a greater chi-squared = 0.0080); Pseudo $R^2$ = 0.0930.

non-response are perfect are not used, leaving just 12 industries represented and 11 industry effects to estimate (with the food industry in the intercept). The coefficient for EEI is − 0.130 with an asymptotic t-ratio of − 1.91, with the probability of a greater absolute value for the z-statistic being 0.057. In this much smaller sample and with all of the available industry effects estimated as well as the effects of EEI and MED2PR, the effect of MED2PR is smaller (the coefficient is − 0.0029 with a z-statistic of − 1.00) and no longer significant.

It is perhaps not surprising that in the relatively small 185 observation sample, with several of the two-digit SIC industries not represented, that the estimable industry effects as a whole are not significant. Table 3.9 provides an alternative way of looking at the data. Just the industry effects that appear, in the complete specification, somewhat significantly different from the other industries' effects are estimated. Since the perfect prediction cases could have occurred by chance given the small sample, one specification leaves them in the intercept while the other does not use them, allowing comparison of the two results. The significance of the effect of EEI on the probability of response is robust to all of the various specifications, with its most statistically significant effect in the specifications of Table 3.9.

To sum up the effect of EEI on the probability of responding, note that the average value of EEI in the 185 company sample is 2.82. Further, companies with large value for EEI result in a large dispersion for the distribution; the estimated standard deviation for EEI is 6.93. Using the probit estimates from the first specification in Table 3.9, consider the probability of responding for any of the industries in the intercept — that is, any industries other than SIC 33, 35, or 38. The probit index is $0.0384 - 0.123*EEI - 0.00393*MED2PR$. If the industry has the average value for EEI and MED2PR, then its probit index = $0.0384 - (0.123)*(2.82) - (0.00393)*(172.06) = - 0.9846558$, with corresponding probability equal to 0.162 taken from the cumulative normal distribution. If the industry has the average value for MED2PR, but has EEI that is a standard deviation above the average for the sample, then its probit index = $0.0384 - (0.123)*(2.82 + 6.93) - (0.00393)*(172.06) = - 1.837046$, with corresponding probability equal to 0.0331. In this case, the probability of responding to the survey falls from about a 16 percent chance of responding to about a 3 percent chance given the increase from the average value of EEI to a value a standard deviation above the mean.

The probability model estimated provides insight about the factors that determine response to the survey, and the model will be used in subsequent chapters to allow a control for selection into the sample when hypotheses about the behavior of companies are tested. As we estimate models throughout the rest of the book, the presence or absence of the response effects will be discussed. Although we shall find a response effect in some of the models estimated subsequently, experimentation shows that the results for the effects of the explanatory variables are essentially unchanged whether or not controls for response are entered into the models. The factors determining response that we have explored in this chapter are nonetheless interesting in themselves. In general, firms were less likely to respond to the survey if they faced greater environmental problems.

# 4. The Industrial R&D Response

This chapter uses the environmental R&D survey to describe industry's R&D response, after the passage of the Clean Air Act Amendments (CAAA) in 1990, to the environmental damage from the production and use of industrial products. The definition of environmental R&D is broad, namely any R&D work that would be aimed at reducing or controlling or limiting the effects of emissions that could potentially damage the environment. The work could be research about emissions themselves, or R&D for new processes that lessen emissions, or R&D for new products that will be cleaner in production or in use.

The responses to the questionnaire reveal that manufacturing companies perceive a large percentage of their performed industrial R&D is related to improving environmental performance of their products and processes. That percentage is estimated to be 23.9 percent for the 71 surveyed manufacturing firms that described their environmental R&D. The percentage and breakdowns of the percentage across different types of firms are developed in this chapter. Also developed in the chapter are descriptions of the manufacturers' research about emissions and of their R&D to improve the environmental performance of products and processes. For all of the various types of environmental R&D, the financing, cooperative activity, appropriability conditions, cost, risk, efficiencies of R&D scale, and government mandates are described and compared.

## THREE TYPES OF ENVIRONMENTAL R&D

Responses show that firms do distinguish types of environmental R&D, although the evidence shows that companies do not always consider those types to be mutually exclusive. Stated differently, companies can allocate their environmental efforts into the three categories distinguished in the survey, but the responses indicate they often consider their environmental R&D an integrated whole, with overlapping efforts in various types of R&D that can be distinguished in principle.

Environmental R&D is a portion of the total R&D performed by a company (but not necessarily financed by the company). Responding companies were asked to define research and development (R&D) expenditures conventionally as in the surveys of industry by the National Science Foundation. The total R&D effort of a responding company includes the company's own R&D — company-financed and performed R&D. It includes as well R&D performed by the company under contract for other firms and the government — R&D financed by others but performed by the company. Stated concisely, the portion of R&D that is environmental R&D is any R&D that has the ultimate goal of improving the environmental performance (reducing hazardous emissions, for example) of processes or products.

The questionnaire gathers information about three basic types of industry's *environmental* R&D aimed at toxic air emissions, although those three types need not be mutually exclusive. First, there is the background research about toxic emissions themselves. The knowledge generated by such research would typically be generic information that could be applied in the development of products and processes. Second, there is R&D investment in new processes, and third R&D for new products. Considering those three types of environmental R&D, the data support the view that the R&D is roughly divided into thirds. About a third of the R&D is devoted to research about emissions, a third to improving the emissions performance of processes, and a third to developing new products with better emissions performance in production or use.

The survey focused on research related to air-emissions. The questions about the proportions of air-emissions research, air-emissions reducing process R&D, and air-emissions reducing product R&D in a company's environmental R&D are:

12. Approximately what part of your company's total research and development effort for environmental projects is for research on toxic air emissions?

25. Approximately what part of your company's total research and development effort for environmental projects is for R&D on new processes to lessen toxic air emissions?

38. Approximately what part of your company's total research and development effort for environmental projects is for R&D on new products to lessen toxic air emissions?

The categories are not in general mutually exclusive; hence, the sum of their shares in environmental R&D need not for a given company equal 1.0 even if 100 percent of its environmental R&D were aimed at air emissions. In principle, one could disentangle two parts of the R&D for a new process or a new product. There would be the part focused solely on studying the properties of the emissions themselves, and then there would be the part focused on the development of the new process or product that would reduce those emissions or other harmful environmental effects. In practice, a company might not be able to disentangle the part of its process or product R&D that was focused solely on studying the properties of the emissions themselves. Thus, some companies will report process or product R&D expenditures not only in the second and third set of questions, but also in the first set of questions that asked about research about the emissions. Some companies will also report R&D expenditures as both product and process R&D. Some research projects will have results for both products and processes, and additionally a company's R&D could produce a new process for the company that would also be a new product for the company because it would be sold to other companies.

In addition to overlapping efforts in background research in emissions and R&D for products and processes, the efforts aimed at air emissions will be intermingled with those aimed at emissions into the land or into water and with environmental problems more generally. Thus, a company might logically report that 100 percent of its environmental R&D effort is concerned with air emissions, even though some of that R&D, perhaps a large part of it, is also concerned with other environmental problems as well.

For all of the foregoing reasons, there is no general expectation about the sum of the shares of the three types of environmental research as revealed in the answers to questions 12, 25, and 38.[1] A company with research only on air emissions could logically report shares summing to anything from one (when the three types of R&D are mutually exclusive) to three (when the three types are completely overlapping). Companies with environmental research applicable to problems other than air emissions could have shares summing to anything from fractions between zero and one to three, depending on the range from complete mutual exclusivity to complete overlapping for the research.

However, expectations not withstanding, the averages of the responses across all the reporting companies do, approximately, add up. There were 42 respondents reporting background research on air emissions in their answers to question 12, and on average they reported 37.4 percent of their environmental R&D in that category. The 58 firms reporting some process R&D for air emissions in their answers to question 25, had 30.3 percent of their environmental R&D there. Finally, 39 companies reported product R&D to reduce air emissions in their answers to question 38. On average, they had 33.8 percent of their environmental R&D in that air-emissions reducing product R&D.

---

[1] Reinforcing that expectation is the fact that the responses were gathered as ranges from which the midpoint was recorded.

The approximate adding up result supports the inference that companies perceive their environmental R&D efforts across types of problems — such as water and land emissions as contrasted with air-emissions — as overlapping. The adding up of the averages across companies suggests that the companies R&D efforts in other areas such as water and land emissions are part of the investments considered to be R&D aimed at air emissions.

Further, the variance because the sums within companies range from fractions to amounts greater than 1.0 supports the inference that companies see the different types of R&D — research about emissions, process R&D, product R&D — for a particular type of problem — such as air emissions — as overlapping.

## ENVIRONMENTAL R&D RELATIVE TO TOTAL R&D

Companies consider much of the R&D they do to be concerned with improving the environmental performance of their products and processes. That fact is implicit in the answers to questions 12 and 13 for companies reporting background research on emissions:

12. Approximately what part of your company's total research and development effort for environmental projects is for research on toxic air emissions?

13. Approximately what part of your company's total research and development effort (for all projects, not just environmental ones) is for research on toxic air emissions?

The answer to question 13 divided by the answer to question 12 yields the company's ratio of environmental R&D to total R&D.

The answer is also implicit in the answers to the questions about process R&D and about product R&D for the companies that reported those types of R&D. The answers will of course be somewhat different, in part because the set of respondents differs.

25. Approximately what part of your company's total research and development effort for environmental projects is for R&D on new processes to lessen toxic air emissions?

26. Approximately what part of your company's total research and development effort (for all projects, not just environmental ones) is for R&D on new processes to lessen toxic air emissions?

38.    Approximately what part of your company's total research and development effort for environmental projects is for R&D on new products to lessen toxic air emissions?

39.    Approximately what part of your company's total research and development effort (for all projects, not just environmental ones) is for R&D on new products to lessen toxic air emissions?

*Table 4.1.  Prediction of Environmental R&D/Total R&D for Companies Reporting R&D on Toxic Air Emissions* *

| Sample | Number of Observations (n) | Predictions ('Forecast'), Mean | Predictions, Standard Deviation | Standard Errors of Forecast, Mean | Standard Errors of Forecast, Standard Deviation |
|---|---|---|---|---|---|
| Background research only | 3 | 0.240 | 0 | 0.148 | 0 |
| Process R&D only | 12 | 0.257 | 0.039 | 0.143 | 0.0027 |
| Product R&D only | 9 | 0.383 | 0.072 | 0.154 | 0.0068 |
| Background & process | 17 | 0.183 | 0.060 | 0.141 | 0.0018 |
| Background & product | 1 | 0.240 | . | 0.148 | . |
| Process & product | 8 | 0.263 | 0.048 | 0.143 | 0.0032 |
| Background, process, & product | 21 | 0.204 | 0.071 | 0.143 | 0.0058 |
| Full sample | 71 | 0.239 | 0.085 | 0.144 | 0.0057 |

*The predictions are estimated well in the sense that the standard error of the forecast is typically considerably smaller than the prediction. For the full sample of 71 observations, the mean value of the ratio of the prediction (or forecast) to the standard error of the forecast is 1.643. The standard deviation for those 71 ratios equals 0.518. The minimum ratio equals 1.074, and the maximum ratio equals 2.988.

One might anticipate differences across the three samples in the perceptions of R&D that is environmentally related.  On the whole, the sample of companies developing new products would be thinking in their R&D about the environmental performance of the products and the processes to make them and the fundamental characteristics of any emissions associated with those products and their manufacture. Such companies would conceivably have more of the R&D problem before them; conceivably, a larger portion of total R&D would be perceived as environmental R&D. Using the data provided by the responding firms, evidently those companies developing new products perceive that a larger proportion of their total R&D effort is environmentally related than the group of companies reporting either process R&D or emissions research but no product R&D.

Table 4.1 shows the prediction for the proportion of total R&D that is taken by environmental R&D for firms reporting emissions research, process R&D, or product R&D in various combinations.

Table 4.1 is derived from the reported data using a simple model shown in Table 4.2. Most companies that had multiple categories of R&D answered the three sets of questions consistently, so that the derived ratio of environmental R&D to total R&D was the same for a respondent regardless of which of the three sets of two questions was used for the derivation. Some had inconsistent answers, however. In all, 47 of the 71 respondents reporting R&D to reduce air-emissions problems had consistent answers across the several ways of calculating the ratio of environmental to total R&D for the company. The variable Envtotot is the consistent answer given for that ratio. The 47 companies with values for Envtotot were used to estimate Table 4.2's model, and it is then used to predict the ratio for all of the companies with various types of R&D and in various industries. In the small sample of 47 firms, only two industry effects were significantly different from the others. Perhaps as one would expect, with the constant term for all of the other industries represented in the sample, the chemicals and petroleum industries have significantly higher ratios of environmental to total R&D. The model also controls for a company's types of environmental R&D, with product R&D in the intercept and the dummy variables DBACKGROUND and DPROCESS indicating the presence of the other two types of environmental R&D.

The predicted, or forecast, ratio of environmental R&D to total R&D is the sum of the estimated intercept term and the products of the estimated coefficients and the values taken on by their associated variables for an observation. That predicted ratio is described in Table 4.1. For comparison with the estimates there, we can note the within sample (that is, the sample of the 47 firms used to estimate the model) predictions for the ratio of environmental to total R&D. For the 47 firms with consistent answers on the survey, the model's predictions of the ratio average 0.228 with standard deviation 0.074.[2] The standard errors of forecast for these 47 firms have mean of 0.144 and standard deviation 0.0055. These results are essentially the same as what the model predicts for the full

---

[2] For the actual values derived from the questionnaire, in the sample of 47, the mean for Envtotot is 0.228 with standard deviation 0.150.

sample;  the 47 firms provide a good representation of all 71 firms in terms of the variables in the model.

Clearly there is some randomness in the assignment by busy executives of answers to difficult numerical questions.  For the present derived result that comes from the ratio of two such answers, formal statistical procedure can help identify the proportion of total R&D devoted to environmental R&D.

*Table 4.2.  OLS Model for Predicting the Ratio of Environmental to Total R&D for Manufacturing Companies with Air Emissions R&D, Dependent Variable Envtotot for the 47 Companies with Consistent Answers*

| Variable | Coefficient | t-statistic with 42 degrees of freedom | Probability > \|t\| |
|---|---|---|---|
| DBACKGROUND | − 0.0954 | − 2.07 | 0.045 |
| DPROCESS | − 0.0897 | − 1.55 | 0.128 |
| D28 Chemicals | 0.136 | 2.49 | 0.017 |
| D29 Petroleum | 0.158 | 1.90 | 0.064 |
| Intercept | 0.336 | 5.68 | 0.000 |

$R^2$ = 0.241;  Adjusted $R^2$ =0.168;  F for the equation as a whole = 3.33 with 4 and 42 degrees of freedom;  probability of a greater F given the null hypothesis that all effects are zero = 0.0186.

One might expect that response bias could affect the foregoing results about the ratio of environmental to total R&D.  However, selection into the sample based on response to the questionnaire does not have a significant effect on the model of the proportion of total industrial R&D devoted to environmental R&D. The specification shown in Table 4.2 was reestimated with the addition of the hazard rate derived from Chapter 3's probit model for response to the survey. The hazard rate controls for the possibility of sample selection bias as explained by Greene (1997, p. 978) and Maddala (1983, pp. 268–9).  The response term is not significant; its coefficient has a t-statistic of only 0.53, and its presence does not have a large effect on the other estimates for the explanatory variables' coefficients.

## AIR-EMISSIONS R&D

If we keep in mind the differences in the samples for each of the types of environmental R&D, and also the fact that the categories are to some extent overlapping, we can in Table 4.3 make a cautious summary using questions 13 and 26 and 39. In particular, as an average percentage of total R&D performed *by those companies that report R&D on toxic air emissions*, the background research on air emissions averages 7.4 percent, process R&D to reduce air emissions averages 7.8 percent, and product R&D to reduce air emissions averages 15.1 percent.

*Table 4.3. Air Emissions R&D as a Percentage of Total R&D Performed: For Firms Reporting R&D to Reduce Toxic Air Emissions*

| Type of Air Emissions R&D | Percentage of a Company's Total R&D |
| --- | --- |
| Background Research | 7.4 (n = 42) |
| Process R&D | 7.8 (n = 58) |
| Product R&D | 15.1 (n = 39) |

While there were 132 companies with substantial manufacturing operations that answered the questionnaire, only 42 report any air-emissions research, 58 any process R&D, and 39 any product R&D. Many of those firms are the same. Among the 132 respondents with substantial manufacturing activity, just 71 of the companies report any type of air-emissions R&D. There are 61 of the 132 respondents that report they do no air-emissions related R&D, 24 respondents reporting just one of the three types of R&D, 26 reporting two of the three types, and 21 reporting all three types of air-emissions related R&D. Table 4.4 shows the exact breakdown for the 71 companies reporting air-emissions R&D, including the breakdown for the subset of 68 firms that report air-emissions research or R&D for Title III chemicals. For the entire 132 respondents — all of which have substantial amounts of total R&D — the average proportion of their total R&D effort taken by air-emissions research is 2.3 percent, by air-emissions process R&D is 3.4 percent, and by air-emissions product R&D is 4.5 percent.

## TITLE III AIR-EMISSIONS REDUCING R&D

The focus of the questionnaire is on industrial research to lessen emissions of Title III air emissions. The appendix to this chapter shows the Title III chemicals addressed by the R&D of 51 of the 68 respondents with Title III R&D, and it shows as well the number of those 51 respondents that indicated research concerned with each particular chemical. Those 51 respondents reported in the optional final question (question # 40) the Title III chemicals of concern in their R&D.

*Table 4.4. Respondents with Air-Emissions R&D*

| Types of R&D | Number of firms with air-emissions R&D | Number of firms with Title III R&D |
|---|---|---|
| All three types | 21 | 21 |
| Background research on air emissions only | 3 | 4* |
| Process R&D on air emissions only | 12 | 12 |
| Product R&D on air emissions only | 9 | 8 |
| Background research and process R&D only | 17 | 15 |
| Background research and product R&D only | 1 | 1 |
| Process R&D and product R&D only | 8 | 7 |
| Total | 71 | 68 |

*Note that the firms in the second column are not necessarily a subset of those in the first column. For example, here one of the four firms has both background research on emissions and process R&D on emissions as well, but the process work is not on Title III emissions while the background research is. Hence the firm is counted in column 2 but not in column 1.

The responses to the survey allow estimates of the proportion of each respondent's total R&D that is aimed at the reduction of Title III emissions. The questions used to obtain an estimate for Title III emissions background research are:

1. Is your company conducting **research on any toxic air emissions** (those on the enclosed list **or** other air toxics)?

2. Approximately what part of your company's research on toxic air emissions focuses on those chemicals listed as hazardous air pollutants in Title III of the Clean Air Act amendments of 1990 (please see the list of Title III pollutants

attached)?

4. Is your company's research on toxic air emissions

COMPANY FINANCED

PERFORMED UNDER CONTRACT FOR ANOTHER COMPANY

GOVERNMENT FINANCED

13.   Approximately what part of your company's total research and development effort (for all projects, not just environmental ones) is for research on toxic air emissions?

The proportion BACK3 of company-financed R&D devoted to background research on Title III toxic air emissions is estimated as zero if the company answered no to question 1. If question 1 was answered yes, then BACK3 was estimated as the product of three things: (1) the answer (as a proportion) to question 2, (2) 1.0 if the answer to question 4 indicated company financed research (as was the case for all respondents who answered yes to question 1), and (3) the answer (as a proportion) to question 13.

Of the 132 manufacturing respondents, 42 answered yes to question 1, and their average proportion in response to question 2 was 0.58, including one firm for which the proportion going to Title III research was zero. That is, all but one of the 42 answering yes to question 1 had some Title III air-emissions research. All 42 indicated in question 4 that the research on air emissions was company financed. One respondent among the 42 also said that there was some research performed under contract for others, and three of the 42 said that there was some government funding of their emissions research. For the 42 firms reporting air-emissions research, their average proportion in response to question 13 was 0.074.

Using the information, we can calculate BACK3. For the 41 firms for which BACK3 is positive, the average value of BACK3 is 0.043. For all 132 respondents, BACK3 averaged 0.013.

The questions used to estimate the effort for Title III process R&D are:

14. Is your company conducting **R&D to develop new processes lessening toxic air emissions** (those on the enclosed list or other air toxics)?

15. Approximately what part of your company's R&D on new processes to lessen toxic air emissions is concerned with those chemicals listed as hazardous air pollutants in Title III of the Clean Air Act Amendments of 1990 (please see

the list of Title III pollutants attached)?

17. Is your company's R&D on new processes to lessen toxic air emissions

COMPANY FINANCED

PERFORMED UNDER CONTRACT FOR ANOTHER COMPANY

GOVERNMENT FINANCED

26.    Approximately what part of your company's total research and development effort (for all projects, not just environmental ones) is for R&D on new processes to lessen toxic air emissions?

The proportion PROC3 of company-financed R&D devoted to process R&D to reduce Title III air emissions is estimated as zero if the company answered no to question 14. If question 14 was answered yes, then PROC3 was estimated as the product of three things: (1) the answer (as a proportion) to question 15, (2) 1.0 if the answer to question 17 included company financed R&D (as was the case for all respondents who answered yes to question 14), and (3) the answer (as a proportion) to question 26.

Of the 132 manufacturing respondents, 58 answered yes to question 14, and their average proportion in response to question 15 was 0.55, including 3 firms for which the proportion of their air-emissions process R&D going to Title III chemicals was zero. That is, all but three of the 58 answering yes to question 14 had some Title III air-emissions process R&D. All 58 indicated in question 17 that the process R&D on air emissions was company financed. None of the respondents among the 58 said that there was some process emissions R&D performed under contract for others, and one of the 58 said that there was some government funding of their process emissions R&D. For the 58 firms reporting air-emissions process R&D, their average proportion in response to question 26 was 0.078.

Using the information, we can calculate PROC3. For the 55 firms for which PROC3 is positive, the average value of PROC3 is 0.042. For all 132 manufacturing respondents, PROC3 averaged 0.018.

The questions used to estimate the effort for Title III product R&D are:

27. Is your company conducting **R&D to develop new products lessening toxic air emissions** (those on the enclosed list or other air toxics)?

28. Approximately what part of your company's R&D on new products to lessen toxic air emissions is concerned with those chemicals listed as hazardous

air pollutants in Title III of the Clean Air Act Amendments of 1990 (please see the list of Title III pollutants attached)?

30. Is your company's R&D on new products to lessen toxic air emissions

COMPANY FINANCED

PERFORMED UNDER CONTRACT FOR ANOTHER COMPANY

GOVERNMENT FINANCED

39. Approximately what part of your company's total research and development effort (for all projects, not just environmental ones) is for R&D on new products to lessen toxic air emissions?

The proportion PROD3 of company-financed R&D devoted to product R&D to reduce Title III air emissions is estimated as zero if the company answered no to question 27. If question 27 was answered yes, then PROD3 was estimated as the product of three things: (1) the answer (as a proportion) to question 28, (2) 1.0 if the answer to question 30 was company financed (as was the case for all respondents who answered yes to question 27), and (3) the answer (as a proportion) to question 39.

Of the 132 manufacturing respondents, 39 answered yes to question 27, and their average proportion in response to question 28 was 0.38, including two of the 39 firms for which that proportion was zero. That is, all but two of the 39 answering yes to question 27 had some Title III air-emissions product R&D. All 39 indicated in question 30 that the product R&D on air emissions was company financed. None of the 39 respondents said that there was some air-emissions product R&D performed under contract for others, and three of the 39 said that there was some government funding of their air-emissions product R&D. For the 39 firms reporting air-emissions product R&D, their average proportion in response to question 39 was 0.15.

Using the information, we can calculate PROD3. For the 37 firms for which PROD3 is positive, the average value of PROD3 is 0.072. For all 132 respondents, PROD3 averaged 0.020.

The questionnaire provides descriptive information about each type of air-emissions R&D. The nature of the research is described, and then there is information for the research about the financing, any cooperative activity, the appropriability conditions, the relevance of appropriability for the R&D investment decision, the costliness, the riskiness, economies of scale, and government mandates. Tables 4.5, 4.6, and 4.7 summarize the information for the firms reporting air emissions R&D.  The descriptive information in the tables reveals several interesting comparisons across the three types of R&D.

*Table 4.5.   Background Research on Air Emissions:    Percentage for Those Responding to the Particular Question\**

---

Description of the research (n = 42)
3.  Is your company's research on toxic air emissions
STUDY OF THE TOXICITY OF THE EMISSIONS 9.5
STUDY OF THE PROCESSES CREATING THE EMISSIONS  78.6
OTHER  35.7

Financing (n = 42)
4.  Is your company's research on toxic air emissions
COMPANY FINANCED  100
PERFORMED UNDER CONTRACT FOR ANOTHER COMPANY  2.4
GOVERNMENT FINANCED  7.1

Cooperative activity (n = 42)
5.  Is your company's research on toxic air emissions
PERFORMED IN A COOPERATIVE VENTURE WITH OTHER FIRMS  26.2
PERFORMED IN A COOPERATIVE VENTURE WITH GOVERNMENT  4.8
PERFORMED INDEPENDENTLY  88.1

Appropriability conditions (n = 42)
6.  How difficult will it be to realize a normal (customarily expected given the riskiness of the projects) return from your company's investments in research on toxic air emissions?  NOT DIFFICULT 11.9  SOMEWHAT DIFFICULT 31.0  DIFFICULT 38.1  VERY DIFFICULT 19.0

Decision-making relevance of appropriability conditions (n = 42)
7.  How important is the ability to realize normal returns for your company's decision to do research on toxic air emissions?  NOT IMPORTANT 9.5   SOMEWHAT IMPORTANT 42.9  IMPORTANT 33.3  VERY IMPORTANT 14.3

Costliness (n = 41)
8.  Compared to your company's other research projects, is research on toxic air emissions LESS COSTLY 31.7  ABOUT AS COSTLY 53.7  MORE COSTLY 14.6

Riskiness (n = 41)
9.  Compared to your company's other research projects, is research on toxic air emissions LESS RISKY 22.0  ABOUT AS RISKY 53.7  MORE RISKY 24.4

Efficiencies (n = 41)
10.  Compared to your company's other research projects, for research on toxic emissions are efficiencies from the size of the research effort   LESS IMPORTANT 29.3   ABOUT AS IMPORTANT 65.9  MORE IMPORTANT 4.9

Government mandate (n = 42)
11.  Is your company's research on toxic air emissions a response to a specific government request for information?      YES 16.7  NO 83.3

---

\*Some questions have answers that are not mutually exclusive, so their totals add to more than 100 percent.  Other questions have answers that are mutually exclusive, yet they may not add to 100 percent because of rounding error.  n = the number of respondents for the question.

*Table 4.6. Process R&D on Air Emissions: Percentage for Those Responding to the Particular Question\**

Description of the research (n = 58)

16. Is your company's R&D on new processes to lessen toxic air emissions for
PROCESS TECHNOLOGY TO BE USED BY YOUR FIRM IN ITS PRODUCTION 93.1
PROCESS TECHNOLOGY TO BE USED BY OTHERS IN THEIR PRODUCTION 12.1
PROCESS TECHNOLOGY EMBODIED IN A PRODUCERS GOOD TO BE SOLD 6.9
OTHER 3.4

Financing (n = 58)

17. Is your company's R&D on new processes to lessen toxic air emissions
COMPANY FINANCED 100
PERFORMED UNDER CONTRACT FOR ANOTHER COMPANY 0
GOVERNMENT FINANCED 1.7

Cooperative activity (n = 58)

18. Is your company's R&D on new processes to lessen toxic air emissions
PERFORMED IN A COOPERATIVE VENTURE WITH OTHER FIRMS 15.5
PERFORMED IN A COOPERATIVE VENTURE WITH GOVERNMENT 1.7
PERFORMED INDEPENDENTLY 93.1

Appropriability conditions (n = 58)

19. How difficult will it be to realize a normal (customarily expected given the riskiness of the projects) return from your company's investments in R&D on new processes to lessen toxic air emissions? NOT DIFFICULT 13.8 SOMEWHAT DIFFICULT 37.9 DIFFICULT 37.9 VERY DIFFICULT 10.3

Decision-making relevance of appropriability conditions (n = 58)

20. How important is the ability to realize normal returns for your company's decision to do R&D on new processes to lessen toxic air emissions? NOT IMPORTANT 20.7 SOMEWHAT IMPORTANT 31.0 IMPORTANT 34.5 VERY IMPORTANT 13.8

Costliness (n = 58)

21. Compared to your company's other R&D projects, is R&D on new processes to lessen toxic air emissions LESS COSTLY 25.9 ABOUT AS COSTLY 53.4 MORE COSTLY 20.7

Riskiness (n = 58)

22. Compared to your company's other R&D projects, is R&D on new processes to lessen toxic air emissions LESS RISKY 24.1 ABOUT AS RISKY 51.7 MORE RISKY 24.1

Efficiencies (n = 58)

23. Compared to your company's other R&D projects, for R&D on new processes to lessen toxic air emissions are efficiencies from the size of the research effort LESS IMPORTANT 27.6 ABOUT AS IMPORTANT 58.6 MORE IMPORTANT 13.8

Government mandate (n = 58)

24. Is your company's R&D on new processes to lessen toxic air emissions a response to a specific government regulation?     YES 51.7 NO 48.3

\*Some questions have answers that are not mutually exclusive, so their totals add to more than 100 percent. Other questions have answers that are mutually exclusive, yet they may not add to 100 percent because of rounding error. n = the number of respondents for the question.

First, the information describes the purposes for the types of R&D. The emissions research is largely for the study of the processes that create the toxic

*Table 4.7. Product R&D on Air Emissions: Percentage for Those Responding to the Particular Question\**

---

Description of the research (n = 39)

29. Is your company's R&D on new products to lessen toxic air emissions for
PRODUCTS PRODUCED WITH CLEANER PROCESS TECHNOLOGY 53.8
PRODUCTS THAT WILL HAVE LOWER TOXIC AIR EMISSIONS WHEN USED 56.4
PROCESS TECHNOLOGY EMBODIED IN A PRODUCERS GOOD TO BE SOLD 5.1
OTHER 10.3

Financing (n = 39)

30. Is your company's R&D on new products to lessen toxic air emissions
COMPANY FINANCED 100
PERFORMED UNDER CONTRACT FOR ANOTHER COMPANY 0
GOVERNMENT FINANCED 7.7

Cooperative activity (n = 39)

31. Is your company's R&D on new products to lessen toxic air emissions
PERFORMED IN A COOPERATIVE VENTURE WITH OTHER FIRMS 20.5
PERFORMED IN A COOPERATIVE VENTURE WITH GOVERNMENT 10.3
PERFORMED INDEPENDENTLY 89.7

Appropriability conditions (n = 39)

32. How difficult will it be to realize a normal (customarily expected given the riskiness of the projects) return from your company's investments in R&D on new products to lessen toxic air emissions? NOT DIFFICULT 17.9 SOMEWHAT DIFFICULT 25.6 DIFFICULT 41.0  DIFFICULT 15.4

Decision-making relevance of appropriability conditions (n = 39)

33. How important is the ability to realize normal returns for your company's decision to do R&D on new products to lessen toxic air emissions? NOT IMPORTANT 2.6 SOMEWHAT IMPORTANT 28.2 IMPORTANT 41.0 VERY IMPORTANT 28.2

Costliness (n = 39)

34. Compared to your company's other R&D projects, is R&D on new products to lessen toxic air emissions LESS COSTLY 7.7 ABOUT AS COSTLY 61.5 MORE COSTLY 30.8

Riskiness (n = 39)

35. Compared to your company's other R&D projects, is R&D on new products to lessen toxic air emissions LESS RISKY 12.8 ABOUT AS RISKY 56.4 MORE RISKY 30.8

Efficiencies (n = 39)

36. Compared to your company's other R&D projects, for R&D on new products to lessen toxic air emissions are efficiencies from the size of the research effort LESS IMPORTANT 12.8 ABOUT AS IMPORTANT 74.4 MORE IMPORTANT 12.8

Government mandate (n = 39)

37. Is your company's R&D on new products to lessen toxic air emissions a response to a specific government regulation?       YES 61.5 NO 38.5

---

\*Some questions have answers that are not mutually exclusive, so their totals add to more than 100 percent. Other questions have answers that are mutually exclusive, yet they may not add to 100 percent because of rounding error. n = the number of respondents for the question.

emissions. The process R&D is overwhelmingly for new process technology to be used by the company in its own production. The product R&D is roughly evenly split between development of new products that will be produced with cleaner processes and new products that will have lower emissions when used.

Regarding the financing of the environmental R&D, all of the respondents report that company funds are invested in each of the three types of R&D. But government funding for the work is clearly much more likely for emissions research and for product R&D than for process R&D. Research is performed under contract for other companies in just the emissions research.

Most air-emissions R&D projects are performed independently by the companies, but cooperation with other companies is substantial, with the largest amount, by the percentage of respondents reporting such activity, of cooperative activity in emissions research. Cooperative activity with the government is greatest for product R&D, yet substantial for emissions research, and considerably less for process research.

The differences in appropriability conditions for the three types of R&D may not be as large as some observers would expect. One would expect that appropriating returns from the background research on emissions would be most difficult. That is the case, but perhaps not to the extent one would expect. Perhaps the research has more proprietary content than generic content given the information that the research is about processes creating emissions and those processes may be somewhat idiosyncratic to the companies. More likely is the possibility that only companies doing the research will have the know-how to benefit from the results, and that is true for the other two categories of research as well. To some extent though expectations about appropriability conditions are confirmed, with about twice the percentage of respondents reporting very difficult appropriability conditions for emissions research than for process R&D. Appropriability difficulties for product R&D are intermediate between the other two types of R&D. They are greater than those for process R&D, with the percentage of companies reporting very difficult conditions in product R&D one and a half times as great as the percentage for process R&D. That percentage reporting very difficult conditions appropriating returns to product R&D is about three-fourths the percentage for the emissions research. These results do conform to expectations since the emissions work would be expected to be more generic, and since cooperative activity and government funding is greatest in emissions and product work, and since overwhelmingly the process work is for the companies' own use.

The decision-making relevance of appropriability conditions is markedly greater for product R&D. Given that the new products developed are to be sold to others, the much larger proportion of respondents reporting that the appropriation of returns from their environmental product R&D is important or very important is expected.

On the whole, environmental R&D is about as costly as other types of R&D projects, although there is a marked tendency for R&D that develops new products with better emissions performance in use or in production to be more costly than the typical R&D project.

Roughly half of the respondents indicate that environmental R&D projects are about as risky as other R&D, although roughly a quarter of the respondents have environmental R&D that is less risky than their other projects and about a quarter report that the environmental investments are more risky. The product R&D, perhaps because it is aimed at new products for sale to others and because companies report that its appropriability conditions are important for decision making, is more likely to be seen as relatively risky. Almost a third of the respondents report that their environmental product R&D is more risky than their typical R&D project.

Most respondents indicate that efficiencies from the size of the R&D effort tend to be neither more nor less important than for other R&D projects, although somewhat more than a quarter of the respondents report that efficiencies are less important in their emissions research and their process R&D to reduce emissions.

The companies report that their background research about emissions is much less likely (than the R&D to develop new processes and products) to be in response to a specific request for information from the government.

In subsequent chapters, the descriptive information that has been reported in this chapter will be used to explore hypotheses about the industrial R&D response to environmental problems that are caused as a byproduct of industrial production.

# 5. A Theoretical Model

This chapter describes a theory of industry's environmental R&D response to public concerns about toxic emissions. Two theoretical models are used to describe the environmental R&D response. First, environmental R&D behavior is analysed with a model in which a company's environmental R&D effort is a function of each explanatory variable's effect on the probability of failure to meet regulatory standards for emissions. Second, the uncertainty about the R&D to improve environmental performance is described in more detail, with the R&D investment both shifting and changing the shape of the probability distribution of outcomes. The two theoretical descriptions of environmental R&D imply that in the context of public concern — expressed through government — about industrial environmental damage, companies invest in environmental R&D. Further, the models imply that those investments are expected to increase as the competitive pressures on the companies increase.

## AN UNCOMPLICATED MODEL OF TITLE III R&D BEHAVIOR

Given the public's concern about Title III pollutants as expressed in the Clean Air Act Amendments of 1990, if a company's industries are associated with Title III pollutants to a sufficient extent, we expect the company is likely to do Title III R&D. It makes those investments to be ready to operate in the anticipated regulatory environment. Is there more to understanding the behavior of the companies that do Title III R&D than knowing that they operate in industries that must deal with Title III pollutants? To offer an answer, consider first a very simple model of the behavior of companies that must deal with the uncertainty of potential regulatory and private liabilities because of Title III air toxic emissions. After presenting the simple model and developing its predictions, a second more complicated model will be presented. We shall see that the predictions of the simple model carry over to the more complicated one. Thus, for those preferring to make models as simple as possible, but not more so, we will have established that the essential predictions necessary for the hypothesis tests are generated by the simpler model as well as the more complex one.

Industrial organization theories of R&D investment typically model races to win a prize or a share of the prize. Scott (1993, Chapter 8) uses simulations to illustrate that approach. Martin (2002) provides a highly developed and articulated model that has evolved in the industrial organization literature from the simplest racing models. For the R&D on toxic emissions, this chapter uses a different — complementary and not inconsistent, but more general in that details of any race are not specified — approach that is tailored to the particular issue of environmental R&D.

The problem of the toxic emissions addressed by the R&D studied in this book is the classical problem of an externality in production or use of a product. New government regulations evolving from the Clean Air Act Amendments of 1990 have provided the impetus for the R&D. Our first model assumes that a firm engages in the environmental R&D to reduce the probability that it will be forced out of business or at best find its value substantially lessened because its production processes or products do not meet the environmental standards imposed by government.

The firm then maximizes its value, and other things being the same that value is a decreasing function of the probability that environmental standards are not met. Thus, assuming for simplicity that a production process or product with environmental quality greater than required has no value to the firm, the firm chooses the environmental R&D to maximize the firm's value over two states — one in which it has met regulatory requirements and one in which it has not. Now, the probability of failing to meet the standards is a function of several things. Those things include the extent of the toxic emissions problem that must be solved, the basic science and applied knowledge about the R&D problem, the technology of conducting R&D in the area, the market structure (broadly defined) that constrains the success of the firm's R&D efforts, and of course the R&D investment itself. In this simple model, the firm has made all of its other decisions — about price, advertising, quality, R&D more generally — to maximize its value given that it meets environmental standards imposed by the government.

We shall show that the first-order condition for value maximization with respect to investment, RIII, in R&D on Title III toxic emissions then implies that the optimal level of such R&D expenditure is a function of the factors exogenous to the firm given the existing technology. Those factors include the extent (numbers and types of chemicals) of the emissions problem, the knowledge base, the technology for R&D, and the market structure. Furthermore, we shall show that the hypotheses, about the signs of the effects for the various variables on the probability of meeting regulatory standards, imply all of the signs for the effects of the exogenous variables on R&D expenditures in the R&D equation derived from the first order condition for those expenditures. Moreover, the hypotheses have implications for policy because the R&D equation shows that public pressure does interact with the profit-maximizing behavior of companies to increase their R&D investments in pollution-reducing processes and products.

Let $V$ represent the value of the company given that the firm meets environmental regulations that evolve in response to Title III of the Clean Air Act Amendments of 1990. Let alpha, $\alpha$, be the probability that the company does not meet those regulations. Alpha is a function of R&D investment; hence, $\alpha = \alpha(RIII)$. The company, assumed to be risk neutral, chooses its Title III R&D, RIII, to maximize:

$$Z = V(1 - \alpha) + \lambda V \alpha - RIII = V(1 + \alpha(\lambda - 1)) - RIII,$$

where $\lambda < 1$ and $\lambda V$ is the value of the firm in the case regulations are not met — assets may have to be liquidated or fines paid and so forth.[1]

The first order condition is:

$$\partial Z / \partial RIII = V(\lambda - 1)(\partial \alpha / \partial RIII) - 1 = 0,$$

because for this simplest model, we assume that V itself does not depend on RIII. Instead, RIII affects $\alpha$, the probability that process technology or product performance will not be sufficiently nonpolluting to allow production and use to meet regulatory muster.

Hence, the first-order condition for maximizing value is:

$$V(\lambda - 1)(\partial \alpha / \partial RIII) - 1 = 0, \text{ and } (\partial \alpha / \partial RIII) = (1 / V(\lambda - 1)) < 0.$$

Let $\alpha = A X_1^{\beta_1} X_2^{\beta_2} \cdots X_n^{\beta_n} RIII^{\beta_R}$, where A is a constant and $X_i$ denotes the $i$th explanatory variable other than the decision variable RIII. The functional form for alpha is a natural one, with the betas representing elasticities of alpha with respect to the explanatory variables. The predictions of the model evolve from the functional form for alpha and the assumption of value-maximizing behavior. Then:

$$(\partial \alpha / \partial RIII) = \beta_R A X_1^{\beta_1} \cdots X_n^{\beta_n} RIII^{\beta_R - 1} = (1 / (V(\lambda - 1)) < 0; \ \beta_R < 0.$$

The value V of a company that meets environmental regulations is a function of the company's profits. Those profits depend on the firm's choices for its

---

[1] Note — in contrast with the literature about crime — there is no distinction in this simple model between violating the regulations and getting caught doing so. That distinction is not important for the generalizations to be developed with the model.

decision variables — price, advertising, and so forth. For our empirical model, we assume that in our sample of firms overall company value for a company meeting regulations is related to the company's observed sales. Then in the sample, let $V = a(SALES)^b$ where $SALES$ denotes a company's sales and where $a$ (and possibly $b$) is industry-specific. Then, we have the estimable model:

$$\ln RIII = \frac{-\ln a - \ln(1-\lambda) - \ln(-\beta_R) - \ln A}{\beta_R - 1}$$

$$+ \left(\frac{-b}{\beta_R - 1}\right) \ln SALES + \sum_i^n \left(\frac{-\beta_i}{\beta_R - 1}\right) \ln X_i .$$

The first term will be captured in the intercept and, if the parameters vary across industries, by industry effects. Interactions of those effects with the explanatory variables can allow industry specific responses to those variables. Thus, the regression model fits industry dummy variables and the natural logarithms of company sales and of the other explanatory variables determining the natural logarithm of Title III R&D for the company except for random error.

The second order conditions are consistent with the presumption that $\beta_R$ is less than zero; and thus, the sign of the coefficient for an explanatory variable $X_i$ is determined by the sign of the elasticity of the probability of failing to meet regulatory mandates with respect to that explanatory variable. The coefficient on an explanatory variable is $\left(-\beta_i /(\beta_R - 1)\right)$, and the second order conditions imply that the sign of the coefficient on an explanatory variable $X_i$ is determined by the sign of $\beta_i$, the elasticity of $\alpha$ with respect to the $i$th explanatory variable. That follows since the second order conditions require, among other things, that $V(\lambda - 1)(\partial^2 \alpha / \partial RIII^2) < 0$, which assuming that $0 < \lambda < 1$ (at worst there is some net salvage value for the assets) implies $(\partial^2 \alpha / \partial RIII^2) > 0$. Now, since $\partial^2 \alpha / \partial RIII^2 = (\beta_R - 1)\beta_R A X_1^{\beta_1} \cdots X_n^{\beta_n} RIII^{\beta_R - 2} > 0$, then $(\beta_R - 1)\beta_R > 0$, which implies $(\beta_R - 1) < 0$, which is consistent with the presumption that $\beta_R < 0$.

For example, we expect that the probability of failing to meet the mandates increases with an increase in Title III pollutants, and then a variable measuring their presence would increase investments in Title III R&D. For SALES, the elasticity of RIII with respect to a company's size is expected to be positive, because the elasticity b of value V with respect to SALES is hypothesized to be positive.

Given the foregoing description of the firm's R&D decision, what would be the effect of an increase in competitive pressures faced by a firm? I shall advance and exposit the following hypothesis: greater *dynamic* competitive pressure is expected to imply that regulatory mandates will be more stringent. The competition is the dynamic competition for new products and processes. In the industrial organization literature there has been a debate about whether the conventional measure of competition — conventionally assumed to decrease with seller concentration — or instead the measure of Schumpeterian competition — assumed to increase with seller concentration — will be associated with dynamic competition to introduce new products or processes (Baldwin and Scott, 1987; Cohen and Levin, 1989; Scherer, 1970; Scherer, 1984). In Chapter 6, our data will allow a test of whether competition matters for environmental R&D and, if so, whether the dynamic competitive pressures that cause higher environmental R&D occur in concentrated or instead relatively unconcentrated markets. Effective *dynamic* competitive pressure is expected to matter, because with more effective competition, the state of the art or best-practice emissions performance is expected to be better. Regulators would expect more from firms regarding their environmental performance. Thus, greater effective dynamic competition implies that other things being the same the probability of failing to meet the regulatory mandates is greater. Hence, the optimal environmental R&D investment is hypothesized to increase with the intensity of dynamic competitive pressures. Whether dynamic competition is a product of more concentrated markets as Schumpeter suggested is an open question and one that we shall address with new empirical work.

## MODELING UNCERTAINTY AND RELATIVE PERFORMANCE

We turn now to developing a more complete model; we 'switch gears' and abandon the simple assumptions of the preceding model. In this section we also switch gears in the sense that we start anew with the notation; so alpha, for example, will be used with a new meaning. The reader should therefore take from the preceding section just the predictions for comparison with the predictions of the new, more complicated model to be developed now.

What the literature has called 'Schumpeterian competition' is assumed to come from firms that dominate their concentrated markets. Schumpeterian competition refers to the *dynamic* competition for new products and processes; such dynamic competition is to be contrasted with *static* competition *given* the existing set of products and processes. With the most perfect, textbook static competition in output markets, price would fall immediately to post-innovation market unit costs, leaving no return to R&D investments. For real-world markets with their range of actual competition, dynamic competition with R&D investments can occur in relatively unconcentrated markets with strong static competition, yet not immediate adjustment of market supply to eliminate profits from innovation. Although more static competition in output markets may imply more pressure to cut costs and therefore *reduce* environmental R&D, it is also possible that it would be associated with more pressure to do R&D to, at least temporarily, do better than the long run competitive equilibrium profits and to ensure survival as the market evolves.

This book will use the more complete model — to be developed in this section — to test the hypothesis that the emissions-reducing R&D investments of US manufacturing firms increase in response to *dynamic* competition in the context of emissions regulation. To test the hypothesis, the primary data about US industrial firms' emissions-reducing R&D investments will be used in a more complete model linking R&D investment to emissions reduction and to the value of the firm. In the uncomplicated model of the preceding section, R&D and other characteristics of a company and the industries in which it operates affect the probability that a firm will fail to meet regulatory mandates. In the present section, uncertainty and relative performance of a firm's R&D investments are modeled more explicitly. The company's R&D and other characteristics affect the probability distribution over the environmental performance achieved by the firm's new processes and products. In the more developed model, the value of a company's R&D increases with the environmental performance it achieves and is also influenced by various company and industry characteristics.

The essential link from the regulations to the value-maximizing behavior of a company is that a company invests in R&D to improve its success in meeting the government's emissions standards. In this second, more detailed model, the value of the R&D increases with the environmental performance of new products and processes. Better technologies have more potential to be royalty-earning licensed technologies, and they are less likely to result in fines or other penalties for environmental damage. Good performance increases the likelihood that the firm's technology will satisfy evolving standards that tighten as emissions characteristics of best-practice technology improve.

The model shows that factors that increase the risk of failing to meet the government's emissions standards cause a company to do more R&D. Factors that increase the value of meeting those standards are hypothesized to imply more R&D as well. Competition is a factor that could affect both the value of emissions reduction and the risk of failing to meet the government's standards — with *opposite* effects on R&D investment. The model sorts out the possible effects, and its estimation in Chapter 6 provides an assessment of the actual overall effect.

The model of this section is more general than the uncomplicated model of the preceding section. First, it introduces uncertainty to the R&D investment's outcome. Second, it considers the outcome of research not in the dichotomous terms of whether or not the regulatory standard is achieved. Instead the outcome is described with an index of environmental performance.

In the more detailed model, improvements in emissions performance of process technology or products affect the company's market value. The company conducts R&D on emissions problems, and the R&D affects the probability distribution for the measure, $x$, of its environmental performance in addressing the regulatory standards proposed by the government.

For the R&D on toxic air emissions, the link from R&D to the probability distribution implies that more R&D, *ceteris paribus*, yields a greater expected value for the index, $x$, of environmental performance. At the same time, conducting more R&D should imply that the company is accepting greater risks; thus, the distribution for the performance index has greater variance as R&D investment increases. Low R&D investment implies that the bulk of the probability is for achieving low performance. As R&D increases, the distribution shifts to the right toward higher performance, becomes more symmetric, and has higher variance.

The regulatory standard — or evolving standard in the case of Title III chemicals — that triggers the R&D could reflect any number of concrete policies regarding, for example, the quality of the air. The measure of performance would be based on the standard. For example, if the standard were that 50 percent of some waste product be eliminated from the production process, removing 25 percent achieves 50 percent of the standard while removing 75 percent achieves 150 percent of the standard. Or, the index of performance could be linked to anticipated health benefits. For example, with the clean up of the waste product might come the expectation of better health for those breathing the air near the factories using the process that emits toxic chemicals. The index of environmental performance could be cast in terms of the health benefits achieved with the reductions in emissions achieved.

In the abstract depiction of the R&D investment outcomes below, the probability distribution over the environmental performance achieved allows for the measure x to range from zero to infinity. That allows a smooth and rather general derivation of the empirical model and the predicted effects of the variables, but more specific distributions that would be tied to particular emissions standards should not change the qualitative predictions of the model. The gamma distribution introduced subsequently should be considered a simplifying approximation.

Now, exactly how does success in environmental performance affect competitiveness and the firm's market value? There are several possibilities and none are excluded from the model. If environmental performance is improved by cost-lowering process innovations, better realizations of $x$ could help a firm vis-à-vis its rivals because, for example, its variable costs are lowered relative to theirs. Even when the improvement is simply from end-of-pipe filtering efficiency, it will have cost advantages and improve competitiveness in the context of the government regulations. There are the revenues from licensing innovations to competitors, and there are the gains from avoiding fines for violating the standards. Of course, the funds being devoted to environmental R&D could have been diverted from more productive uses (such as other kinds of R&D which could also enhance market value). We cannot in general be certain that the *social* benefits of the regulatory-induced R&D will exceed their opportunity costs. However, the point of the model is that the regulation causes firms to find it in their *private* interest to invest in R&D to reduce emissions that the government wanted to reduce. The regulation occurred because the governmental process decided to mandate emission reduction.

The variance property described above may seem unusual, so a brief digression to discuss it is in order. For some models of R&D behavior, additional R&D could imply that companies are working their way down the list of projects ranked by their expected payoffs. In those cases, as more projects with lower returns are added to the portfolio, risk could decrease. However, in the present case, the model captures the decision to make bold, qualitatively new, investment decisions at an extensive margin. Rather than being comfortable with a known process and technique for cleaning up and low-risk R&D to clean up at the margin, the company accepts the risks of R&D to find new methods. I want to model an extensive margin here, because I want to capture the idea that when the government announces that it will set standards, it causes firms to undertake a new class of R&D investment. However, as comparison of this section's model with the uncomplicated model in the preceding section will show, the estimating equations and the expected signs for the coefficients of the regressors are the same when variance is ignored and just the expected value of R&D is modeled.

A distribution that has the desired properties for modeling the risk and performance of investments in environmental R&D is the gamma distribution. With the random variable $x$ denoting the measure of environmental performance achieved, we have the probability density:

$$f(x;\ \alpha,\beta)\ =\ \frac{1}{\alpha!\beta^{\alpha+1}}\,x^{\alpha}e^{-x/\beta} \qquad 0 < x < \infty$$

$$=\ 0 \qquad\qquad\qquad \text{elsewhere} \ .$$

In this application, alpha is a nonnegative integer, and beta must be positive.[2] As $\alpha$ increases the probability distribution changes shape — with probability weight shifting (from predominantly lower performance) toward the center of the

---

[2] Note again that, for this second model, all of the notation is defined anew. Thus alpha in this new model is not the same thing as the alpha in the simple, introductory model of the preceding section.

distribution and with the distribution having more symmetry with greater dispersion around a higher central value. Mood and Graybill (1963, p. 127) provide a graphical illustration. Thus as alpha increases, the distribution becomes less skewed and shifts to the right centering over higher performance for the innovative technology and exhibiting greater dispersion.[3]

In the model, $\alpha$ is determined by the company's R&D, by the nature of the toxic emissions problem, and even by the extent of competition — whether associated with lower seller concentration or instead higher concentration and the R&D competition among large firms dominating their markets as Schumpeter hypothesized. Baldwin and Scott (1987, pp. 1–4) provides a review of Schumpeter's writings about technological progress that results given the competition among large corporations that dominate their markets. To survive, the large corporation must innovate, and it is uniquely suited, in Schumpeter's hypothesis, to develop and introduce new processes and products. Thus, although at first glance it must seem somewhat of an oxymoron, we have Schumpeterian competition when a market is concentrated. Unlike conventional discussions of static competition in which competition decreases with seller concentration, with the *dynamic* competition of R&D rivalry, Schumpeter expected competition to increase with seller concentration. Chapter 6 will ask if seller concentration affects environmental R&D.

The presence of competition as a determinant of $\alpha$ may seem surprising; the reason is that with greater R&D competition, both the average innovation is expected to be better and the regulatory standard is expected to be tougher, because the dynamic competition will push the state of the art technology to higher levels and allow regulators to expect more from the regulated companies. So, more dynamic competition implies tougher standards for performance. Environmental performance achieved by a company is, *ceteris paribus*, lower, being judged against a tougher standard and having faced a tougher R&D problem. With greater dynamic competition, the state of the art process and product is better, so the known approaches are better and the next generation of innovations pose a tougher R&D problem. Regulators anticipate that and pose tougher standards when R&D competition is more pronounced.

The basic idea is that more dynamic competition results in better processes and products and higher goals for standards and puts the distribution for

---

[3] Thus, higher R&D results in higher likelihood of larger 'technology jumps.' Higher R&D can have two effects on innovation possibilities. First, the step size of innovation increases — finding new methods and not just cleaning up at the margin. In this case risk (the dispersion around the measure of central tendency) increases with more R&D as captured by the properties of the gamma distribution. Second, higher R&D can lead to an increase in the probability of finding an innovation (big or small). Therefore the uncertainty about whether *any* innovation is found decreases with higher R&D, and this too is captured by the gamma distribution because as R&D increases it shifts rightward over higher performance outcomes at the same time that it exhibits greater dispersion around the higher mean.

environmental performance achieved in its less favorable form, *ceteris paribus*. With a better process and product technology as the state of the art starting point and with the tougher standards, more R&D will be required to achieve a favorable distribution. More R&D competition is hypothesized to create a more difficult R&D problem to be solved, because the average innovation will be better and because a tougher standard is expected from the government. There may of course be spillovers of knowledge among the firms, but the average innovation will be better, so better relative performance is harder to achieve.

Competitors would believe that the greater the dynamic competition, the tougher would be the R&D problem they must solve. The innovation target is a moving one, and the essence of more competition is that competitors anticipate that they must come up with something even better to preempt the best of the innovations introduced by their rivals. In the context of the present paper, the R&D problem can be tougher as dynamic competition increases not only because the R&D rivalry increases the achievement of the average innovation, but also because the regulators may require better performance in an industry with more active R&D.

Thus, $\alpha = \hat{\alpha}$, with $\hat{\alpha}(z)$ denoting the integer nearest to $z$:

$$z = \ln A + \sum_{i}^{n} \beta_i \ln X_i + \beta_R \ln RIII$$

where A is a constant and the $X_i$ denote the various determinants of $\alpha$ other than *RIII*, the company's Title III related R&D. For the gamma distribution, there is, in addition to $\alpha$, the parameter $\beta$ which is the scaling factor that given any alpha, and hence given the shape of the distribution, changes the scale on the two axes (see Mood and Graybill, 1963, pp. 126–7). Beta, the appropriate scaling factor, will be different for different pollution problems, is essentially not of great economic interest, and is determined by idiosyncratic technological characteristics of the toxic emissions problems and the R&D efforts to solve them.

Market value increases with the environmental performance of the innovations achieved, but at a decreasing rate. With extraordinary high performance, there is less value for an increment to that performance than would be the case for low performance. The value of environmental performance is modeled as:

$$V(x) = \gamma - \gamma e^{-x/\theta}$$

where *given* the firm's choices for its decision variables other than RIII, in our sample:

$$\gamma = a(SALES)^b Z_1^{b_1} Z_2^{b_2} \cdots Z_m^{b_m}.$$

Now, the expected value of R&D is:

$$E = \int_0^\infty (\gamma - \gamma e^{-x/\theta})(\frac{1}{\alpha! \beta^{\alpha+1}} x^\alpha e^{-x/\beta}) dx.$$

As we show next, for integer values of $\alpha$, the integral for $E$ reduces to:

$$E = \gamma(1 - \left(\frac{\theta}{\theta+\beta}\right)^{\alpha+1}).$$

Or, simplifying, when the two scaling factors, $\theta$ and $\beta$, are equal:

$$E = \gamma(1 - \left(\frac{1}{2}\right)^{\alpha+1}).$$

The integral giving the expected value of R&D is reduced to the foregoing two equations as follows. The expected value of R&D is:

$$E = \int_0^\infty (\gamma - \gamma e^{-x/\theta})(\frac{1}{\alpha! \beta^{\alpha+1}} x^\alpha e^{-x/\beta}) dx.$$

Simple applications of change of variables as well as integration by parts shows that for integer values of $\alpha$, the integral for $E$ reduces to:

$$E = \gamma(1 - \left(\frac{\theta}{\theta+\beta}\right)^{\alpha+1}) = , \text{ when } \theta = \beta, \ \gamma(1 - \left(\frac{1}{2}\right)^{\alpha+1}).$$

We can rewrite E as:

$$E = \int_0^\infty (\gamma)(\frac{1}{\alpha! \beta^{\alpha+1}} x^\alpha e^{-x/\beta}) dx$$

$$- \int_0^\infty (\gamma e^{-x/\theta})(\frac{1}{\alpha! \beta^{\alpha+1}} x^\alpha e^{-x/\beta}) dx$$

Then, the integrals are solved by substituting $y$ for $x/\beta$ in the first integral and $u$ for $x\left(\frac{\theta+\beta}{\theta\beta}\right)$ in the second, and then using the fact (from Mood and Graybill, 1963, p. 127) that with $\alpha > 0$ integration by parts shows that $\int_0^\infty y^\alpha e^{-y} dy$ equals $\alpha!$. Subtracting $E(\alpha)$ from $E(\alpha+1)$ and simplifying, we have the result that:

$$\left(\frac{\Delta E}{\Delta \alpha}\right) = \gamma\left(\frac{\theta}{\theta+\beta}\right)^{\alpha+1}\left(\frac{\beta}{\theta+\beta}\right) = , \text{ when } \theta = \beta, \ \gamma\left(\frac{1}{2}\right)^{\alpha+2}.$$

The company chooses its Title III R&D, $RIII$, to maximize the net expected value of R&D, $E - RIII$. Maximizing the net expected value of R&D requires that:

$$\frac{\Delta E}{\Delta \alpha}\frac{\Delta \alpha}{\Delta RIII} \approx (\gamma(\frac{\theta}{\theta+\beta})^{\alpha+1}(\frac{\beta}{\theta+\beta}))(\frac{\beta_R}{RIII}) = 1,$$

or for the case where the scaling factors are equal:

$$\frac{\Delta E}{\Delta \alpha}\frac{\Delta \alpha}{\Delta RIII} \approx (\gamma(\frac{1}{2})^{\alpha+2})(\frac{\beta_R}{RIII}) = 1,$$

where $\Delta\alpha/\Delta RIII$ is approximated with $\partial z/\partial RIII$.

With no loss in the generality of our interpretations to come, but with the gain of considerably less clutter in the expressions, I shall use the case for which the scaling factors are equal. It is important to have developed the more general case, since in fact the scaling factors would be likely to differ, but equalizing them will reveal most clearly, with a minimum of algebraic detail, the essential theory linking the explanatory variables to R&D investment. Thus, choosing R&D to maximize net expected value requires that:

$$\ln \gamma + (\alpha + 2)\ln(\tfrac{1}{2}) + \ln \beta_R - \ln RIII = 0.$$

Approximating $\alpha$ with $z$, we have for the first order condition:

$$\ln a + b \ln SALES + \sum_i^m b_i \ln Z_i + 2\ln(\tfrac{1}{2})$$
$$+ \ln(\tfrac{1}{2})\Big(\ln A + \sum_i^n \beta_i \ln X_i + \beta_R \ln RIII\Big) + \ln \beta_R - \ln RIII = 0$$

The second order condition (needed to establish the signs on the model's coefficients) then requires:

$$(1 / RIII)\big((\ln(\tfrac{1}{2})\beta_R) - 1\big) < 0$$

which implies that $\big((\ln(\tfrac{1}{2})\beta_R) - 1\big) < 0$.

From the first order condition, we have a model that can be estimated and that predicts the signs for the explanatory variables' coefficients:

$$\ln RIII = \left(\frac{1}{(1 - (\beta_R \ln(\tfrac{1}{2})))}\right)\left(\begin{array}{l} \ln a + 2\ln(\tfrac{1}{2}) + \ln(\tfrac{1}{2})\ln A + \ln \beta_R + b\ln SALES \\ + \sum_i^m b_i \ln Z_i + \sum_i^n (\ln(\tfrac{1}{2}))\beta_i \ln X_i \end{array}\right).$$

Note that the coefficient on a value-shifting variable such as $\ln SALES$ or $\ln Z_i$ is going to have the sign predicted for the elasticity ($b$ or $b_i$) of $\gamma$ with respect to that variable, because the estimated coefficient is of the form $\big(b_i / (1 - (\beta_R \ln(\tfrac{1}{2})))\big)$ where $(1 - (\beta_R \ln(\tfrac{1}{2})))$ is greater than zero from the second order condition. For example, since we believe that firms with larger sales will derive greater value from an innovation, we expect that the estimated coefficient on $\ln SALES$ will be positive.

The interpretation of the coefficients estimated for the probability-shifting variables that determine $\alpha$ is more complicated. The regression of $\ln RIII$ on

$\ln X_i$ gives coefficients of the form $\left( \beta_i \ln(\frac{1}{2}) / (1 - (\beta_R \ln(\frac{1}{2}))) \right)$ where $(1 - (\beta_R \ln(\frac{1}{2})))$ is greater than zero from the second order condition and of course $\ln(\frac{1}{2})$ is negative. Thus, if the estimated coefficient is less than zero, the underlying $\beta_i$ is greater than zero, while if the estimated coefficient is greater than zero, then the underlying $\beta_i$ is negative. So, for example, a variable that is hypothesized to make $\alpha$ smaller — such as more competition because it causes the regulators to choose a tougher R&D problem or more generally because given competition each competitor expects that it will need to solve a tougher R&D problem to remain competitive — has a negative $\beta_i$. So, the expected sign is positive for the coefficient from the regression of $\ln RIII$ on the natural logarithm of a competition index. Greater dynamic competition results, *ceteris paribus*, in a less favorable distribution for the index of environmental performance achieved, because results are judged against a tougher standard. What would have been good performance with less competition looks less than good; and therefore, competition stimulates a company's R&D to make distribution of environmental performance more favorable. At the level of R&D investment for which marginal benefit would equal marginal cost if competition were less intense, the company would perceive that the marginal benefit of more R&D would exceed the marginal costs.

The intercept term captures several parameters that would be expected to vary by industry category; the effects of these parameters can be controlled for with dummy variables to capture the industry effects. Note also that a variable that appears as both a $Z_i$ and an $X_i$ in the basic theoretical model will have an estimated coefficient in the empirical model that combines the variable's effect on value with its effect on the shape and location of the probability distribution. The next chapter estimates the model, and the estimated coefficients of the regressors are interpreted with such combined effects in mind.

# 6. The Hypothesis Tests

The models in the preceding chapter have described how economic incentives link various characteristics of companies and their industries to investments in environmental industrial R&D. In the context of public concern about industrial environmental damage, those characteristics are the determinants of environmental R&D. The models — one relatively uncomplicated and the other with more detail about the uncertainty and relative performance of a company's environmental R&D — yield the same expectations about the impact of the determinants on R&D investments.

This chapter will estimate the effects of those determinants, and thereby test some hypotheses.[1] The presence of toxic emissions problems of concern to the public is expected to affect the probability and amount of environmental R&D to address those problems. Yet there are other important determinants of environmental R&D. A key hypothesis emerging from the theory in Chapter 5 is that dynamic competitive pressures increase environmental R&D investments. The test of the hypothesis will not only allow assessment of whether it has any explanatory power. Further, if competition is important, the test will show whether the type of competition that counts is the 'Schumpeterian' competition from large firms in concentrated markets or instead dynamic competition emerging in unconcentrated industries and in industries facing import competition.

In terms of the simpler of the two models in Chapter 5, the estimation will ask if variables increasing the probability that a company will fail to meet future

---

[1] A preliminary look at the survey data appeared in Scott (1996, 1997). In this chapter, we are able to look at the data in new ways. First, all of the data are now available for analysis, whereas the earlier studies had only the process R&D data to examine. Second, the earlier studies examined just the companies that had Title III R&D, but in this chapter we use the Tobit model to study all of the companies. The explanatory variables prove to be important for the decision to invest in Title III R&D as well as for the amount of R&D conditional on such investment being made. Further, the smaller sample sizes available for the preliminary studies dictated a simple log-log specification. In the context of the Tobit model, the nonlinearities in the variables can be described with a second-order Taylor's series approximation that reveals much more about the relationships than the log-log functional forms used in the earlier studies.

emissions standards also increase the amount of R&D the company devotes to meeting the challenge. In terms of the more complicated model, the question is whether variables that lower a firm's relative environmental performance (shift leftward the distribution for the relative performance of a company's environmental R&D) will increase its investments in that R&D. Both models hypothesize the same effects on environmental R&D for the various characteristics of firms and industries, although the issue of what type of competition matters is an open one.

The estimation of the model provides answers to questions of interest for technology policy regarding toxic air emissions. Does the government's commitment to standards increase R&D investments designed to reduce emissions? The model addresses that question. First, innovations are hypothesized to be more valuable because they are needed to meet the standards. Second, firms are hypothesized to increase R&D to reduce the risk of failing to meet the standard or, more generally, having relatively poor performance. Does domestic competition increase or decrease emissions-reducing R&D? It might decrease the value of innovation because profits in the post-innovation market may be reduced by imitation or competing substitute processes or products. However, it may increase the risk of failing to meet the government's emissions standards if competition improves the state of the art, best practice emissions and thus allows tougher regulatory standards. Does international competition increase R&D — perhaps because of anticipation of bigger markets and more value for an innovation, or perhaps because of inducing (by influencing the toughness of the standard) a shift in the probability distribution for the percentage of regulatory standards achieved? Or will it decrease R&D effort because companies simply give up in the face of foreign competitors who do not have to meet the same standards? Will private R&D be stimulated by cooperative R&D ventures? Will it be stimulated by the size of the firm? These and other questions are answered by estimating the model with the data gathered.

Of the 150 respondents to the survey described in Chapter 3, 132 had manufacturing operations; this chapter examines their R&D behavior. As seen in Chapter 4, of those 132, 68 reported research and development work related to Title III emissions, with one or more of the three types of R&D — emissions research, process R&D, or product R&D — aimed at reducing emissions of Title III chemicals.

An appropriate procedure for analysing the determinants of the presence of company R&D aimed at Title III chemicals is Tobit analysis (Maddala, 1983, Chapter 6; Greene, 1997, pp. 962–5). With the estimates of the parameters provided in this chapter, the interested reader can compute a variety of expectations tailored to specific settings for the explanatory variables. We shall begin by describing how to compute those expected values for Title III R&D and for the marginal effects of the explanatory variables, and then we shall present the estimated parameters of the model.

The Tobit model assumes that for the dependent variable $y_i$ and the fixed $1 \times k$ vector $x_i$ of explanatory variables, and an index function $y_i^* = x_i\beta + \varepsilon_i$, then $y_i = y^*$ if $y^*$ exceeds 0, while $y_i = 0$ if $y_i^* \leq 0$. $\beta$ is a $k \times 1$ vector of unknown parameters, while $\varepsilon_i$ are independently and normally distributed errors with expected value of zero and homoskedastic variance $\sigma^2$. For the present problem, $y_i$ is the observation of Title III research on emissions or R&D on processes or products — RIII in the theory of Chapter 5 — for the $i$th of the 132 responding companies with manufacturing operations. The Tobit model estimates $\beta$ and $\sigma^2$ using the 132 observations on RIII (some at zero and some greater than zero) and the explanatory variables.

Because the Tobit model will examine companies with no Title III R&D as well as those with R&D, we shall not estimate the model in the log-log form used for the functional forms explicating the theory with a minimum of notation in Chapter 5. In our actual estimation, we shall capture the nonlinearities in the model with quadratic specifications that are linear in the parameters estimated although nonlinear in the variables. That approach amounts to a second order Taylor's series approximation with zero cross-partial derivatives (see Greene, 1997, pp. 50–51).

Dividing each term of the model through by $\sigma$ to give normalized coefficients, one can compute the standard normal index value for an observation, and the value of the cumulative normal distribution for that index value corresponds to the probability of observing Title III process R&D for the company. The non-normalized Tobit coefficients themselves are predicting the value of RIII (that prediction for the observed RIII is zero when the regression model yields a non-positive value for the dependent variable).

Using the Tobit model, we can combine the information about the coefficients and about the probability of doing Title III process R&D and estimate the expected value of RIII given any specified values for the explanatory variables $x_i$. As Maddala (1983, pp. 158–9) explains, with $\phi_i$ denoting the density function of the standard normal distribution evaluated at $z_i = x_i\beta/\sigma$, and $\Phi_i$ denoting the cumulative distribution function also evaluated at $z_i = x_i\beta/\sigma$, we have for the expected amount of RIII given that R&D is done: $E(RIII_i \mid RIII_i > 0) = x_i\beta + \sigma(\phi_i/\Phi_i)$. Stated differently, it is, for a given set of explanatory variables, the expected value of the index function $y^*$ given that $y^*$ is greater than zero. Or, equivalently, it is, for those values of the explanatory variables, the expected value of y given that $y^*$ exceeds zero. Thus, it is the mean of the positive ys. The expected amount of RIII for a company randomly drawn from the population and therefore that may or may not do R&D is

$E(RIII_i) = \Phi_i x_i \beta + \sigma \phi_i$. That is the mean of all of the ys, positive or zero. A third prediction from the model is the mean of the index, or latent variable, y* for any given set of explanatory variables, which is simply $E(y_i^*) = x_i \beta$.

Maddala (1983, p. 160) and Greene (1997, pp. 963–4) provide a full development of the marginal effects. First, the derivative of the index function with respect to the set of explanatory variables is of course simply the vector $\beta$. For $y_i$ the marginal effect is scaled by the probability $\Phi_i$ since $\partial E[y_i \mid x_i] / \partial x_i = \beta \Phi_i$. Finally, for the marginal effects given that y is positive, $\partial E(y_i \mid y_i^* > 0) / \partial x = \beta[1 - z_i(\phi_i / \Phi_i) - (\phi_i / \Phi_i)^2]$.

For changes in the explanatory variables, the unconditional marginal effect on $y_i$ can be decomposed into two parts (Greene, 1997, p. 964). The derivative of the expected value with respect to the explanatory variables equals the sum of two terms. First, there is the probability of $y_i$ greater than zero multiplied by the change in the expected value of $y_i$ given that $y_i$ is greater than zero. Second, there is the change in the probability that $y_i$ is greater than zero multiplied by the expected value of $y_i$ given that $y_i$ is greater than zero. 'Thus, a change in $x_i$ has two effects. It affects the conditional mean of $y_i^*$ in the positive part of the distribution, and it affects the probability that the observation will fall in that part of the distribution.' (Greene, 1997, p. 964)

As the foregoing explication of the predictions from the Tobit model should suggest, it will be useful to simulate those predictions for the models that we estimate. That is because visualizing the effects is otherwise difficult given all of the parameters (other than the estimated coefficients for the explanatory variables and the settings for those variables) that affect the predictions. After estimating the model, we shall illustrate the partial effects with simulations of the effects of variables of interest, while holding constant the effects of the other variables when they are at their average values. Using the Tobit model to estimate the determinants of Title III R&D, we can show that companies with operations involving Title III chemicals have responded to public concern about Title III pollutants and invested in R&D to solve the problems. The Tobit estimations show as well that there is more to understanding the behavior of the companies that do Title III R&D than knowing that they operate in industries that must deal with Title III pollutants. Chapter 5's model — of the behavior of companies that must deal with the uncertainty of potential regulatory and private liabilities because of Title III air toxic emissions — points up the hypothesis that greater dynamic competition will increase Title III R&D. Estimation of the Tobit model provides information about that hypothesis.

As explained in Chapter 4, there are 68 firms in the sample with Title III R&D. There are 21 with background, process, and product Title III work, four with background research only, 12 with process R&D only, eight with product R&D only, 15 with background and process work, one with background and product work, and seven with process and product Title III R&D.

We turn now to the estimation of the model for the set of 132 respondents with manufacturing operations, including the 41 firms reporting background research on Title III emissions, the 55 respondents that report Title III process R&D, and the 37 respondents that report Title III product R&D. Given Chapter 5's theory, two sets of interesting candidates for important regressors come readily to mind — namely, the extent of the Title III emissions problems and, additionally, the extent of competitive pressures.

To measure the extent of a company's Title III emissions problems, a new variable NTAPC is introduced. A key factor that distinguishes the 68 companies with Title III R&D from the other respondents is NTAPC, the average value of NTAP for the company's four-digit manufacturing industries.[2] Recall that NTAP is the number of Title III toxic air pollutants associated with a particular four-digit SIC industry. Thus, the presence of Title III emissions problems associated with a company, as indicated by NTAPC, is expected to be an important explanatory factor for R&D investments aimed at Title III problems.

The measures of the pressures of competition are the US national concentration ratio and import competition for each market. The concentration ratio of the

---

[2] Simple counts of chemicals may be most appropriate as contrasted with some sort of attempt to weight their importance. We know the chemicals are dangerous to some degree, but it is — just as with margarine versus butter, etc. — difficult to know exactly how dangerous and how important cleanup for a particular chemical really is. Thus, counts may be better than tons of pollutants. If we wanted solely to compare the problems from one chemical across industries, yes, tons would help. But we are comparing different chemicals as well, and a small amount of one may be more of a problem than a large amount of another. And, for the foregoing reason, counts may be better than sales weighted emissions problems if we are trying to judge the social need for cleanup. On the other hand, the sales weighting might improve the fit of the private R&D decision as a function of the size of the pollution problem regarding developing new processes to solve the emissions problems. As we shall see, the count variable has some explanatory power with the expected sign, but its importance in explaining the company financed air toxic related R&D would probably be greater if, in addition to controlling for a company's total sales and its count variables for pollutants, we could control also for the sales actually associated with each pollutant. Regarding the industry average of NTAP for each company to derive NTAPC, essentially what I am doing is weighting each of the company's industries equally, as if each had equal sales, and then I have controlled for the sales of the company as a whole separately. Thus, the averaging is a way of weighting by sales. I am saying, then, that not all of the company's sales are from products that cause all of the pollutants. In the absence of reliable sales figures by industry and of information about the potential liability from each set of chemicals associated with one of a company's industries, I am essentially assuming that re both factors each industry's chemicals for the several industries in which a company operates should get equal weight. In the search for reliable sales figures, I took a look at the COMPUSTAT business segment data. There are problems with those data for my present purposes. Companies self-report the division of their activities, and the categories can be rather broad and uninformative. The COMPUSTAT analysts then choose one or at most two SIC four-digit industries to represent the sales associated with each of a company's self-reported business segments. The match between the list of industries thereby obtained and what the companies actually do in terms of four-digit industries is not very good.

value of industry shipments is CR4, the four-firm seller concentration ratio as a percentage.[3] A more concentrated market is one with more *Schumpeterian* competition but less competition in the traditional sense of industrial organization. Our hypothesis test in this chapter will let us see which if either concept of competition is competition that is associated with dynamic competition that causes substantial R&D investment. The measure of import competition is IMPS, the ratio of imports to shipments for each four-digit manufacturing industry.[4] Then, for each company — analogous to the way in which NTAPC is derived from NTAP — CR4C and IMPSC are respectively the average values of CR4 and IMPS across the four-digit manufacturing industries in which a company operates.

Why should we expect the SIC four-digit industries to give a reasonable categorization for Schumpeterian competition? Schumpeter (1942) referred to the competition of new products, such as plastics replacing steel, and new processes. The variant of Schumpeterian competition here involves firms being driven to design and adopt newer, more efficient, methods of pollution abatement (new products or new processes with better emissions performance) for fear that if they do not those techniques will be adopted by others.[5] The regulatory authorities would then require similar adoption throughout the 'industry,' to the competitive disadvantage of those behind the curve. Such competition is initially competition among, for example, all firms using soft coal, whether producing electric power or aluminum, or all firms using similar solvents for cleaning machines or tools, whether automobiles or generators. Firms in different product industries may use the same types of energy or similar processes more generally; firms in the same product industry may use different processes. There is also another variant of Schumpeterian competition in the model in the sense that the

---

[3] The measures CR4 are taken from US Department of Commerce (1992, Table 4, pp. 6.4–6.45).

[4] The measure IMPS is constructed from information in US Department of Labor (1991) and in US Department of Commerce (1990). To get imports for each four-digit manufacturing SIC industry using the new 1987 SIC industries, I began with US Department of Labor (1991, Table 1, pp. 3–8). This was the first issue of *Trade & Employment* that used the new 1987 industry categories. Although the issue focused on 1990, import data on the new SIC basis was also given for 1988. The import data are often for combinations of the 1987 four-digit industries, or for pieces of those industries. I used US Department of Commerce (1990, Table 1, pp. 2.5–2.32) to get the value of product shipments for SIC four-digit manufacturing industries and for the five-digit parts of those four-digit industries. These shipments were then used to allocate the imports among the SIC four-digit industries. For example, if an input total was given for two or more four-digit industries, the imports were allocated to the industries in the proportion of those industries' shipments. In many cases, the imports figure was for a combination of five-digit industries. For those cases, the five-digit shipments were used to allocate the imports among the five-digit industries. After all such imputations were made, the imports for the five-digit industries were then combined into their appropriate four-digit industries to get the final figure for imports for the four-digit industry.

[5] The argument assumes, realistically for the innovations of interest, that secrecy, and knowhow that must be practiced to be used effectively, or patents when secrecy will not work, effectively protect the intellectual property of innovators.

latecomers may be disadvantaged in their product markets by their tardiness. The manufacturing companies in our sample are diversified, and we know that those using similar inputs and processes often diversify both their products and their R&D into the same set of industries (Scott, 1993). Companies using similar inputs and processes are especially likely to diversify into the same set of industries if the competition from new products and processes to meet environmental standards is important. Hence, the average measure of seller concentration CR4C for a company's lines of business will to some extent capture both sources of competitive forces — competition from companies using similar inputs and processes and competition from companies in the product markets.

For each company, size is measured with SALES — the sales for the company's most recent fiscal year as of May 18, 1993, in millions of dollars, as reported in *Business Week* (1993). SALES is contemporaneous with the R&D expenses also reported there.

Title III R&D — RIII in Chapter 5's theory — takes three forms — background research on Title III emissions BACK3RD, Title III process R&D PROC3RD, and Title III product R&D PROD3RD. Each of these three variables is an estimate of company-financed R&D stated in millions of dollars, and each is approximated as the product of the company's R&D expenses RD, which are stated in millions of dollars in *Business Week* (1993), and the proportion of the company's total R&D that is for each type of research.[6] We have the proportion for background research on Title III emissions, the proportion of R&D on new processes to lessen Title III toxic air emissions, and the proportion of R&D on new products to lessen Title III emissions. As explained in Chapter 4, each of the proportions is derived from three items on survey's questionnaire.

We also know from Chapter 4's discussion of the responses to the survey that all of the environmental research used own-company financing. But from the responses we also know whether the company did have any funds from other firms or from the government to support its background research, process R&D, or product R&D aimed at emissions problems. The specifications exploring the three types of R&D will use the additional information about those additional sources of financing of the R&D. BACKOTHER takes the value 1 if the company's background research on emissions received additional financing from other companies or from the government, and it takes the value 0 otherwise. PROCOTHER and PRODOTHER are defined analogously and respectively for the company's environmental process and product R&D.

---

[6] Except for random error, a company's company-financed environmental R&D of a particular type is assumed to take the same proportion of the company's company-financed R&D as the company's total environmental R&D effort of that type takes in the company's total R&D effort.

One might expect that investment in one type of emissions research or development would complement investment in another. Background research on emissions might make investment in process R&D or product R&D more productive, and so forth. A straightforward way to ask if such potential for complementarities affect investment is to add qualitative variables for the presence of the other types of R&D in each of the specifications, with DBACKGROUND, DPROCESS, and DPRODUCT denoting the presence respectively of background research, process R&D, and product R&D to address emissions problems.

*Table 6.1. Tobit model of BACK3RD, the Background Research on Title III Emissions**

| Variable | Coefficient | t-statistic | Probability > |t| |
|---|---|---|---|
| NTAPC | 0.370 | 2.39 | 0.019 |
| $NTAPC^2$ | $-0.00274$ | $-1.92$ | 0.057 |
| SALES | 0.000817 | 2.99 | 0.003 |
| $SALES^2$ | $-8.63 \times 10^{-9}$ | $-1.92$ | 0.058 |
| CR4C | $-0.652$ | $-2.04$ | 0.044 |
| $CR4C^2$ | 0.00952 | 2.49 | 0.014 |
| IMPSC | 45.9 | 2.43 | 0.017 |
| $IMPSC^2$ | $-49.9$ | $-2.06$ | 0.041 |
| BACKOTHER | 10.0 | 2.67 | 0.009 |
| Hazard Rate | 0.732 | 0.15 | 0.883 |
| DPROCESS | 10.1 | 4.78 | 0.000 |
| DPRODUCT | 0.258 | 0.14 | 0.886 |
| D28 Chemicals | $-4.00$ | $-1.47$ | 0.144 |
| D29 Petroleum | $-8.03$ | $-1.53$ | 0.129 |
| D35 Industrial Machinery | $-9.10$ | $-2.22$ | 0.028 |
| D36 Electronics | $-4.23$ | $-1.38$ | 0.169 |
| Constant | $-14.6$ | $-1.54$ | 0.125 |

*Number of observations = 132. Log likelihood = $-149.44$. Likelihood Ratio Chi-squared with 16 degrees of freedom = 113.7, with the probability of a greater chi-squared = 0.0000. Pseudo $R^2$ = 0.276. Tobin's $\sigma$, the ancillary parameter for the Tobit model, is estimated to be 5.81 with standard error equal to 0.657. There were 91 left-censored observations at BACK3RD = 0, and there were 41 uncensored observations where BACK3RD > 0.

Estimation of the model shows that the presence of Title III emissions problems and the presence of competitive pressure are associated with greater R&D expenditures to reduce toxic air emissions in manufacturing processes. Further, the competition that counts is not unambiguously either the Schumpeterian view or the conventional industrial organization view of competition that increases as seller concentration falls and import competition increases.[7]

Tables 6.1 through 6.3 present in turn the estimation of the Tobit model for the three types of Title III R&D. The results are broadly similar across the different types of Title III R&D. The effects are nonlinear, with the direction of the effects depending on the levels of the settings for the explanatory variables. Not only do the settings for the variables affect in a nonlinear way the amount of R&D conditional on R&D investment, but they affect the probability of such investment. Because it is difficult to visualize completely the effects of the variables when one simply looks at the Tobit coefficients, after presenting the Tobit estimations, we shall simulate them for interesting settings of the explanatory variables.

Before presenting the estimates and then the simulations to reveal the nature of the nonlinear effects, a simple summary of the findings will serve to introduce them and point up their broad outline across the three types of Title III R&D. The overview of Tables 6.1 through 6.3 is as follows:

Except over very high levels of pollutants, environmental R&D to address Title III pollutants increases with the importance of Title III pollutants in a company's operations. The simulations of the model show that all three types of R&D increase throughout the range of the actual values of NTAPC in the sample of 132 firms until very high levels of pollutants are reached. However, at very high levels of Title III chemicals, the predicted amount of research falls. That may result because the exceptionally high numbers of pollutants correspond to cases where there are many Title III chemicals that industry anticipates will not be greatly affected by new regulations because industry is already handling the chemicals in a satisfactory way.

Larger companies do more Title III R&D. The simulations of the models show that each of the three types of R&D increase with company size throughout most of the range of company sizes in the sample, although R&D declines somewhat for very large company sizes.

Title III R&D is least at intermediate levels of seller concentration. The higher levels of concentration are expected to correspond to greater amounts of Schumpeterian competition while the lower levels correspond to the conventional

---

[7] The earlier, preliminary look at the data (Scott, 1996, 1997) used a simpler log-log functional form that revealed the effect of Schumpeterian competition described below, but was not a rich enough specification to show the effects of both the Schumpeterian view and the conventional view. There were not enough data available for those studies to estimate the more complicated model. With all of the data in hand for all three types of R&D and for all respondents rather than just for those reporting Title III process R&D, the estimations here can use the Tobit model and all observations and a second order Taylor's series approximation to the functional form to reveal the nonlinearities in the effects of all of the variables. With the more comprehensive look at the data thereby allowed, the conclusion about the type of competition — Schumpeterian or conventional — that counts is much richer than could be described in the small data set initially available.

*Table 6.2. Tobit model of PROC3RD, the Process R&D to Reduce Title III Emissions\**

| Variable | Coefficient | t-statistic | Probability > \|t\| |
|---|---|---|---|
| NTAPC | 0.378 | 3.04 | 0.003 |
| $NTAPC^2$ | − 0.00255 | − 2.19 | 0.031 |
| SALES | 0.000667 | 2.93 | 0.004 |
| $SALES^2$ | $4.88 \times 10^{-10}$ | 0.13 | 0.901 |
| CR4C | − 0.742 | − 2.71 | 0.008 |
| $CR4C^2$ | 0.00883 | 2.62 | 0.010 |
| IMPSC | 7.20 | 0.39 | 0.695 |
| $IMPSC^2$ | − 16.4 | − 0.68 | 0.500 |
| PROCOTHER | − 6.34 | − 1.04 | 0.301 |
| Hazard Rate | −7.37 | − 0.86 | 0.391 |
| DBACKGROUND | 7.77 | 4.79 | 0.000 |
| DPRODUCT | 1.86 | 1.25 | 0.215 |
| D20 | 5.40 | 1.34 | 0.182 |
| D25 | 5.09 | 1.17 | 0.243 |
| D26 | 10.7 | 2.48 | 0.015 |
| D29 | − 9.91 | − 2.19 | 0.031 |
| D30 | 10.2 | 1.45 | 0.149 |
| D33 | 7.27 | 1.97 | 0.051 |
| D34 | 5.12 | 1.15 | 0.253 |
| D35 | 12.0 | 1.82 | 0.072 |
| D36 | 12.1 | 2.29 | 0.024 |
| D37 | 12.9 | 3.19 | 0.002 |
| D38 | 8.38 | 1.88 | 0.062 |
| D39 | 13.1 | 1.72 | 0.089 |
| Constant | 0.482 | 0.04 | 0.965 |

\*Number of observations = 132. Log likelihood = − 194.04. Likelihood Ratio Chi-squared with 24 degrees of freedom = 121.16, with the probability of a greater chi-squared = 0.0000. Pseudo $R^2$ = 0.238. Tobin's $\sigma$, the ancillary parameter for the Tobit model, is estimated to be 5.41 with standard error equal to 0.523. There were 77 left-censored observations at PROC3RD = 0, and there were 55 uncensored observations where PROC3RD > 0.

notion of competition. Therefore, the dynamic competition that spurs more R&D does not always originate in solely the 'Schumpeterian' market structures. The most pronounced positive effects of seller concentration on environmental R&D are, however, for the more concentrated, 'Schumpeterian' market structures.

Greater import competition decreases Title III product R&D, and decreases background emissions research above intermediate levels of import competition. There does not appear to be a significant effect of import competition on process R&D.

*Table 6.3. Tobit model of PROD3RD, the Product R&D to Reduce Title III Emissions* *

| Variable | Coefficient | t-statistic | Probability > \|t\| |
|---|---|---|---|
| NTAPC | 3.31 | 4.17 | 0.000 |
| NTAPC$^2$ | $-0.0226$ | $-2.95$ | 0.004 |
| SALES | 0.00458 | 4.07 | 0.000 |
| SALES$^2$ | $-5.07 \times 10^{-8}$ | $-2.61$ | 0.010 |
| CR4C | $-11.8$ | $-7.66$ | 0.000 |
| CR4C$^2$ | 0.153 | 8.37 | 0.000 |
| IMPSC | $-26.7$ | $-0.21$ | 0.834 |
| IMPSC$^2$ | $-164.0$ | $-0.73$ | 0.468 |
| PRODOTHER | 31.1 | 1.73 | 0.087 |
| Hazard Rate | $-141.4$ | $-2.12$ | 0.036 |
| DBACKGROUND | 12.3 | 1.28 | 0.204 |
| DPROCESS | 21.1 | 2.23 | 0.028 |
| D20 | 39.1 | 1.61 | 0.111 |
| D22 | 72.2 | 1.56 | 0.121 |
| D25 | 52.7 | 2.15 | 0.034 |
| D26 | 63.0 | 2.00 | 0.048 |
| D30 | 100.6 | 2.05 | 0.043 |
| D34 | 87.1 | 2.91 | 0.004 |
| D35 | 133.0 | 2.69 | 0.008 |
| D36 | 77.9 | 2.04 | 0.044 |
| D37 | 86.1 | 3.04 | 0.003 |
| D38 | 100.8 | 2.90 | 0.004 |
| Constant | 212.0 | 2.76 | 0.007 |

*Number of observations = 132. Log likelihood = $-195.61$. Likelihood Ratio Chi-squared with 22 degrees of freedom = 98.18, with the probability of a greater chi-squared = 0.0000. Pseudo $R^2$ = 0.201. Tobin's $\sigma$, the ancillary parameter for the Tobit model, is estimated to be 27.14 with standard error equal to 3.23. There were 95 left-censored observations at PROD3RD = 0, and there were 37 uncensored observations where PROD3RD > 0.

The presence of outside financing is associated with more company Title III background research and more Title III product R&D.

If the models of response and R&D expenditures are complete, and if the errors in the response model and the R&D model are correlated, then the expected value of the underlying indicator variable for R&D expenditure, conditional on response to the survey, differs from its unconditional expected value by a function of the probit index from the response model (Maddala, 1983, pp. 268–9).

From the earlier discussion of the Tobit model, Title III R&D, RIII, is observed when the underlying indicator variable — the effect $x_i\beta$ of the explanatory variables plus random error $\varepsilon_i$ — exceeds zero. Given the selection into the sample of 132 manufacturers, following the argument in Maddala (1983, pp. 268–9), conditional on response, the random error $\varepsilon_i$ has two components. There are a random error $\upsilon_i$ and a component $\omega_i$ that depends on the correlation of the errors in the response and R&D models, the standard deviations of those errors, and the outcome for the error in the response model.

Thus, positive RIII is observed for the $i^{th}$ company when $x_i\beta + \omega_i + \upsilon_i > 0$. To estimate the Tobit model with selection into the sample by response to the survey, I assume that, for the 132 manufacturing firms that responded to my survey, $\omega_i$ can be approximated by its expected value $E(\omega_i)$. Hence, positive RIII is observed when $x_i\beta + E(\omega_i) + \upsilon_i > 0$. $E(\omega_i)$ is the expected value of the error in the underlying R&D indicator equation given that the error in the probit equation falls in the critical region such that the company responds to the survey.

The Tobit model is then applied to an augmented set of explanatory variables — the original set plus the appropriate hazard rate. The product of the hazard rate and its estimated coefficient in the Tobit model is an estimate of $E(\omega_i)$. The hazard rate (from Chapter 3's probit model for response to the survey) to control for sample selection is insignificant in the models of background research and process R&D, although it is significant in the specification for product R&D. It has a large effect in the process R&D and product R&D models, even though the effect is statistically significant in only the product R&D model. Thus, for the process and product R&D models, the correlation in the disturbances of the selection model and the R&D model appears to be important. From Chapter 3 we know that companies with fewer environmental problems, other things held equal, were more likely to respond. If one believes that the Tobit models are reasonably complete, for process and product R&D, it appears that there is a downward response bias picked up in the error term of the Tobit model. I shall return to the alternative possibility at the end of this chapter.

The models are exploratory and descriptive, and I cannot claim to be certain that they are complete, 'true' models with joint normal distribution of their errors. The exact and conventional procedure of control for sample selection is sensitive to the assumptions for the true models (Greene, 1997, p. 978; Maddala, 1983, pp. 268–9). For that reason, other functions of the probit index than the hazard rate might be better controls for response. The alternative of entering as an explanatory variable the predicted probability of response leaves essentially the same story about the effects of the explanatory variables.[8] The coefficients and their standard errors are essentially the same for the explanatory variables. For a very few of the fixed industry effects, the coefficient changes size but never sign. The only major difference in using the probability rather than the hazard rate is

---

[8] The robustness of the results is reassuring given some evidence 'that the normal selection-bias adjustment is quite sensitive to departures from normality.' (Maddala, 1983, p. 269)

seen in the model's constant term and the coefficient for the response control. Apart from the constant, the response control itself, and just those few fixed effects that change size somewhat but never direction of the effect, the coefficients and their t-statistics for the entire set of explanatory variables remains essentially the same using the alternative control for response.

Further, if the errors in the response model and the R&D model are correlated, the variance of the underlying indicator variable conditional on response is heteroskedastic. StataCorp (2001, Volume 4, pp. 174–88) provides a generalization, 'intreg,' of the Tobit model that allows computation of robust standard errors for the estimated coefficients — the errors in the observations are allowed to have different variances (heteroskedasticity is assumed). The intreg estimation without the robust option yields the same results as the Tobit models; intreg with the robust option yields the same coefficients, and the significance of the coefficients in the Tobit models presented here remains.

Finally, the estimated coefficients and their significance for the explanatory variables other than industry effects are very robust to dropping the response control from the specifications altogether. For the model of BACK3RD, there is essentially no change in the estimated coefficients and their significance for not only the explanatory variables but for the industry effects as well. For the model of PROC3RD, the constant term changes substantially when the hazard rate is dropped, essentially absorbing the effect that the hazard rate had been capturing. Otherwise, the results are essentially the same. Even for the model of PROD3RD where there is a large statistically significant effect for the hazard rate, only the fixed industry effects and the constant term show marked changes. For the explanatory variables apart from the industry effects, the estimated coefficients and their significance tell essentially the same story with or without the control for response. Its presence in the specification does pick up effects that would otherwise be captured in the constant term and the industry effects, but whether or not it is included we have the same story about the extent of the pollution problem, company size, seller concentration, import competition, outside financing, and other types of environmental R&D.

Industry effects are often discernable, even if not always fully significant by conventional standards. To produce the specifications in the tables for each type of R&D, first a specification with all industry effects was fitted. Second, industry effects that had t-statistics less than 1.0 were dropped, leaving the specifications shown in the tables. The specifications with all industry effects invariably have very similar results for the variables shown in the tables. The specifications in the tables highlight the discernable industry effects, leaving the industries that are not distinguished as different in the intercept. Industry effects might capture, among other things discussed in Chapter 5, differences across industries of the effects of spillovers of research or of innovative output.[9]

We turn now to simulations of the estimated models in order to reveal the nonlinearities in the effects of the variables. We shall simulate the effects of each variable, using the fully specified models of Tables 6.1, 6.2, and 6.3 and given that other variables are held constant. In particular, the tables below will simulate the effects on BACK3RD, PROC3RD, and PROD3RD by computing the Tobit models' predictions of the unconditional expectation of the particular type of Title III R&D as either NTAPC, SALES, CR4C, or IMPSC vary throughout their range in the sample.

*Table 6.4. Means for Simulations*

| Variable | Number of observations | Mean | Standard Deviation |
|----------|------------------------|------|--------------------|
| NTAPC | 132 | 30.80 | 19.47 |
| SALES | 132 | 2721.09 | 7721.90 |
| CR4C | 132 | 36.05 | 11.27 |
| IMPSC | 132 | 0.227 | 0.296 |

Except for the one of those variables that is being varied in each simulation, the other variables are held constant. The remaining three of the four variables of interest for the simulations will be held constant at their means, with the exception of NTAPC. When the effect of one of the other variables is simulated, NTAPC will be held constant at a standard deviation above its mean. That is because we want to illustrate the Title III R&D response of a company facing significant Title III problems. We simulate the effects of the other variables given the presence of substantial Title III R&D.

Further, the estimated effect of the hazard rate for selection is not included in the simulations, although the estimated model being simulated does have the hazard rate in the specification. The response variable captures error that is the

---

[9] From Martin (2002) and the literature that he cites, we know that such spillovers might increase or decrease R&D effort. For that reason, we might not expect measures of spillovers to have discernable effects. Scott (1996, 1997) provides an exploratory look at the measure SPILLC (defined in the Glossary for the variables) of the potential for spillovers of environmental R&D using just the subset of firms with positive process R&D from the survey and the simple log-log specification. Here, in the context of the full sample, and the more complete Taylor's series expansion and Tobit model that the full sample allows, the spillovers variable is not an important factor in the models of environmental R&D.

result of observing only part of the sample and in that case the effect of the response variable is implied by the correlation of the errors for models of response and R&D investment.

The response term has picked up a random effect because of correlation in the errors of the response model and the R&D investment model. With an incomplete model, the response variable could instead pick up an actual systematic effect in the R&D equation. Such a possibility is described by Maddala (1983, pp. 269–70) in the context of a concern that the selection term would pick up nonlinear effects of the explanatory variables in the R&D equation. The solution suggested by Maddala is to include the nonlinear terms in the R&D equation if that makes economic sense, and then ask if the response term is significant. Our models do include the nonlinear terms, and in some cases the response term is still significant.

*Figure 6.1. Unconditional Expected Value of BACK3RD as a Function of NTAPC*

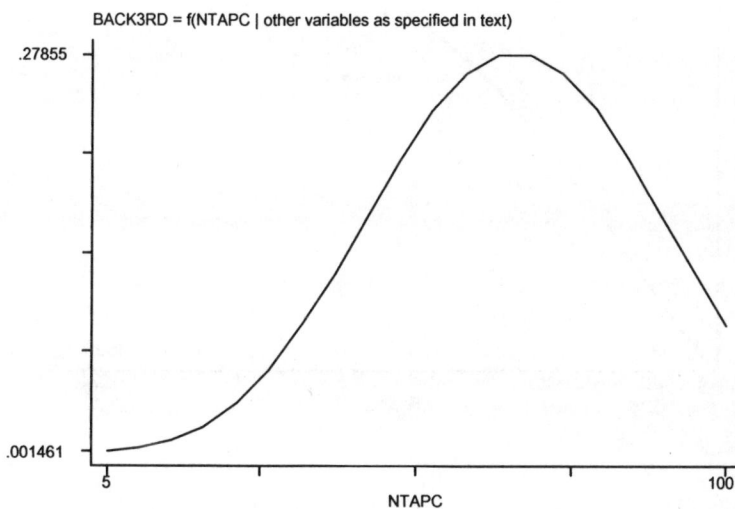

Also, the simulations will assume that there is no outside finance, no other type of Title III R&D other than the one being simulated (so the qualitative variables for the other types are turned off). Note that the presence of other types of Title III R&D will increase the expected amount of the Title III R&D being modeled by several million dollars. Thus, our initial simulations will show lower amounts of the particular type of Title III R&D than would be the case for firms doing all three or two of the types of Title III R&D. For comparison, we simulate the effect of doing the other types of Title III R&D at the end of this chapter. Finally, the simulations also assume that the company has its primary activity in electronics. Thus, D36, the dummy variable for the SIC two-digit industry of electronics is the only one of the industry dummies distinguished in the particular Tobit model that is turned on.

The means and standard deviations for the variables of interest are shown in Table 6.4. When held constant in the simulations, SALES, CR4C, and IMPSC are set equal to their means. NTAPC will be set at 50.3 — a standard deviation above its mean — when it is held constant.

*Figure 6.2. Unconditional Expected Value of PROC3RD as a Function of NTAPC*

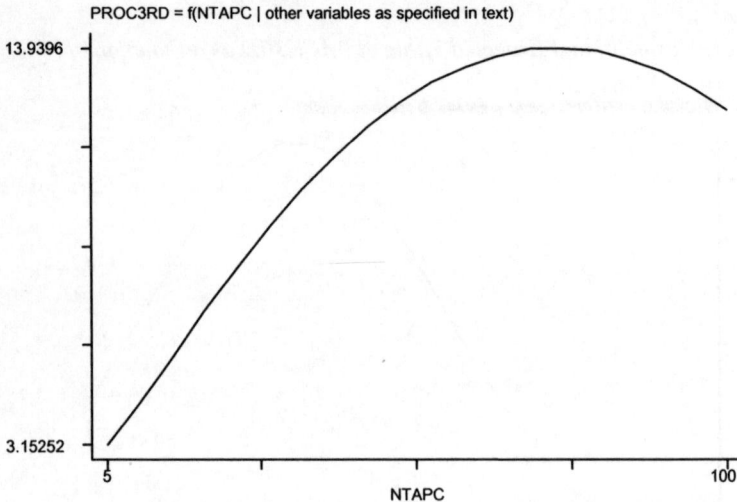

Figures 6.1, 6.2, and 6.3 show the simulations of Title III R&D as NTAPC varies over the range of its values within the sample, holding constant all other variables as specified in the discussion above. The expected amount of Title III R&D increases until the presence of Title III pollutants in a company's production processes reach a level well above the sample mean, and then the R&D declines for very high levels of pollutants. As explained above, the decline for the very high levels of Title III chemicals may reflect chemicals that will not be affected by new regulations because industry already handles the chemicals in a satisfactory way.

Figures 6.4, 6.5, and 6.6 show respectively the simulations for BACK3RD, PROC3RD, and PROD3RD as functions of SALES, holding constant the other variables as specified above, as SALES varies over the range in the sample. All three types of R&D increase with company size until a company's sales are quite large, and then they decline somewhat for BACK3RD and PROD3RD. PROC3RD increases linearly with sales. Clearly for Title III R&D and the range of company sales covering the large firms in the sample, the elasticity of R&D does not exceed 1.0.[10]

*Figure 6.3. Unconditional Expected Value of PROD3RD as a Function of NTAPC*

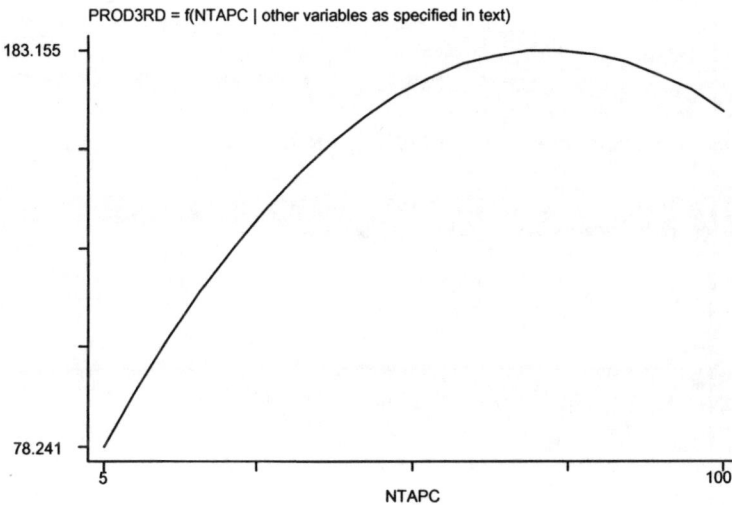

PROD3RD = f(NTAPC | other variables as specified in text)

[10] See Kohn and Scott (1982) for the theory that links the elasticity of R&D with respect to firm size to the literature about Schumpeterian competition.

*Figure 6.4.  Unconditional Expected Value of BACK3RD as a Function of SALES*

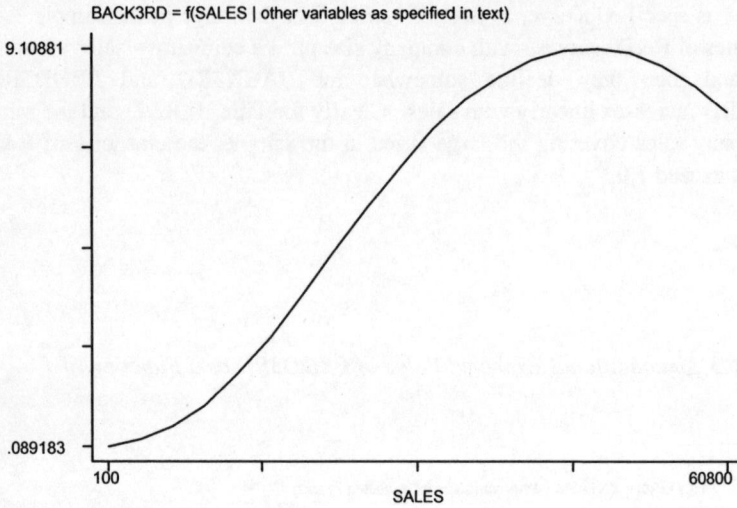

*Figure 6.5.  Unconditional Expected Value of PROC3RD as a Function of SALES*

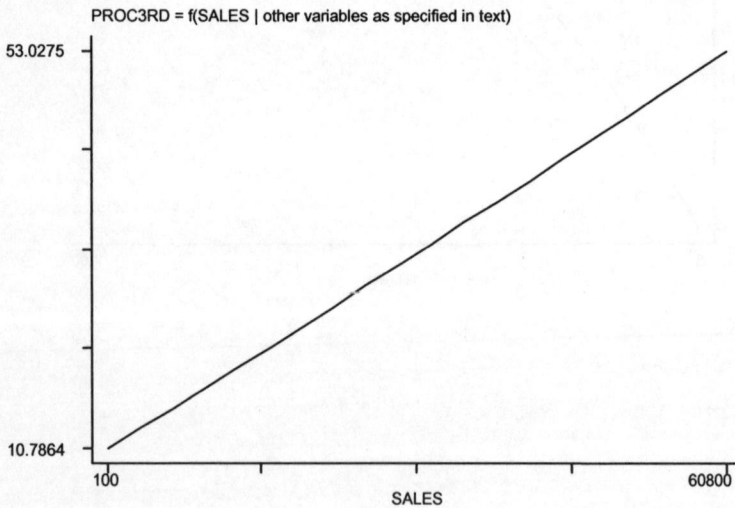

*Figure 6.6. Unconditional Expected Value of PROD3RD as a Function of SALES*

PROD3RD = f(SALES | other variables as specified in text)

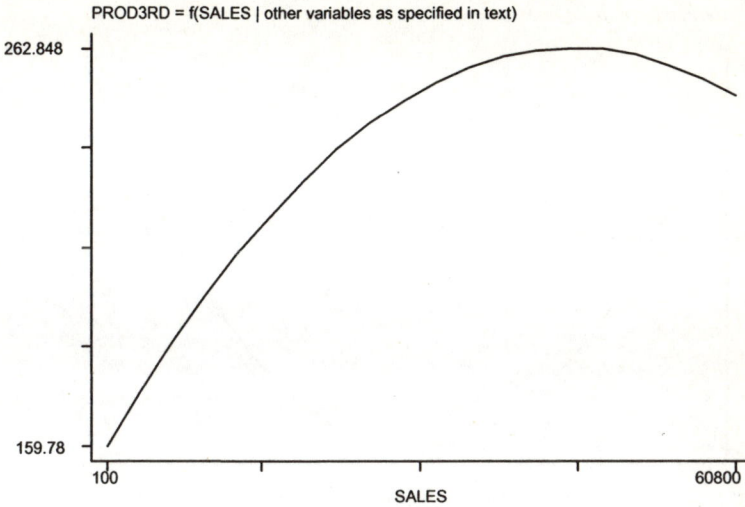

*Figure 6.7. Unconditional Expected Value of BACK3RD as a Function of CR4C*

BACK3RD = f(CR4C | other variables as specified in text)

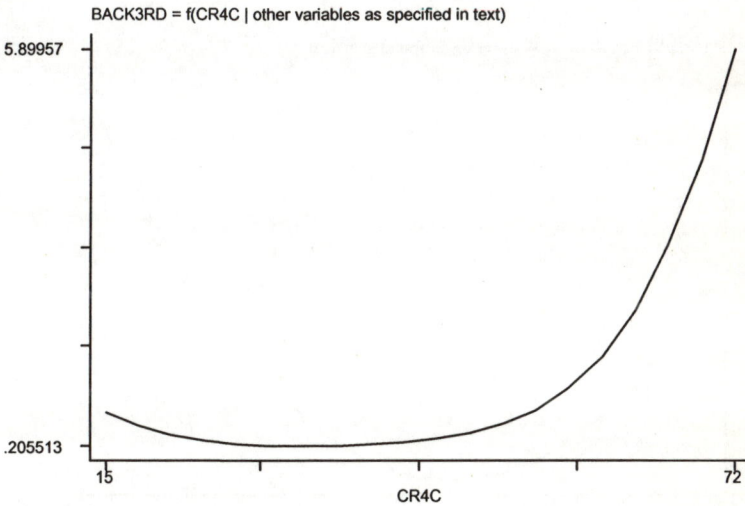

*Figure 6.8. Unconditional Expected Value of PROC3RD as a Function of CR4C*

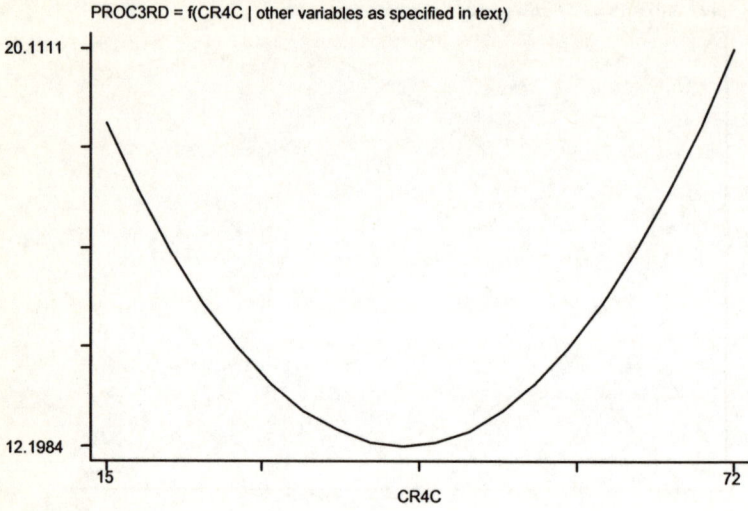

PROC3RD = f(CR4C | other variables as specified in text)

*Figure 6.9. Unconditional Expected Value of PROD3RD as a Function of CR4C*

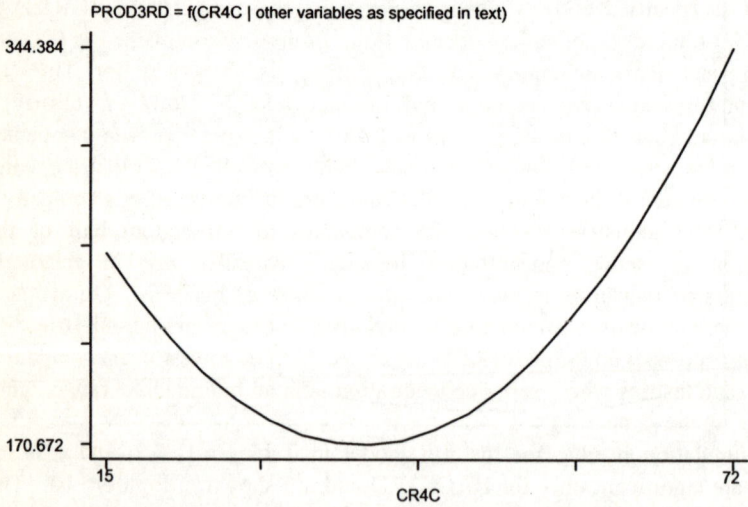

PROD3RD = f(CR4C | other variables as specified in text)

*Figure 6.10. Unconditional Expected Value of BACK3RD as a Function of IMPSC*

BACK3RD = f(IMPSC | other variables as specified in text)

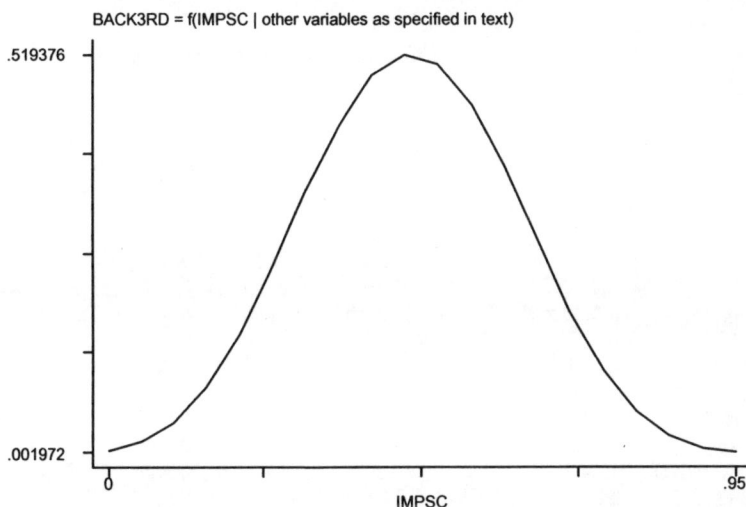

Figures 6.7, 6.8, and 6.9 depict the simulations of Title III R&D as CR4C is varied throughout its range for the sample given the other variables are held constant as specified above. Other things being the same, Title III R&D is greatest for high levels of seller concentration. In the sense explained in Chapter 5, it appears that 'Schumpeterian competition' is important for Title III background research and process and product R&D. However, clearly a substantial amount of Title III process and product is expected when companies operate in industries with low seller concentration as well. Recall that the seller concentration here is the average of seller concentration across all of a company's four-digit SIC industries. Thus, the companies in the bottom half of the distribution of seller concentration here are typically *not* in relatively unconcentrated industries in *every* one of their lines of business. On average, though, these companies *are* operating in relatively unconcentrated industries. In the sample, product and process R&D are expected to be lowest when companies operate in industries where seller concentration is in an intermediate range, other things being the same.

The simulation results, for the full model in Tables 6.1, 6.2, and 6.3, for IMPSC are significant only for BACK3RD and are shown in Figure 6.10. The background research on Title III emissions problems increases with import competition up to an intermediate level of IMPSC, but then BACK3RD falls as the import competition faced by a company increases, other things held constant as specified in the discussion above.

*Table 6.5. The Simple Linear Effect of IMPSC on the Tobit Index for PROD3RD\**

| Variable | Coefficient | t-statistic | Probability > \|t\| |
|---|---|---|---|
| NTAPC | 3.20 | 4.13 | 0.000 |
| NTAPC$^2$ | − 0.0220 | − 2.90 | 0.004 |
| SALES | 0.00477 | 4.33 | 0.000 |
| SALES$^2$ | − 5.30 x 10$^{-8}$ | − 2.76 | 0.007 |
| CR4C | − 11.8 | − 7.70 | 0.000 |
| CR4C$^2$ | 0.154 | 8.40 | 0.000 |
| IMPSC | − 114.5 | − 2.10 | 0.038 |
| PRODOTHER | 31.6 | 1.75 | 0.083 |
| Hazard Rate | − 140.4 | − 2.12 | 0.037 |
| DBACKGROUND | 12.4 | 1.29 | 0.200 |
| DPROCESS | 21.3 | 2.23 | 0.028 |
| D20 | 33.6 | 1.45 | 0.149 |
| D22 | 70.5 | 1.53 | 0.128 |
| D25 | 52.9 | 2.14 | 0.035 |
| D26 | 60.5 | 1.93 | 0.056 |
| D30 | 99.7 | 2.04 | 0.044 |
| D34 | 86.4 | 2.89 | 0.005 |
| D35 | 131.9 | 2.68 | 0.009 |
| D36 | 79.2 | 2.09 | 0.039 |
| D37 | 85.9 | 3.04 | 0.003 |
| D38 | 98.9 | 2.86 | 0.005 |
| Constant | 222.3 | 2.96 | 0.004 |

*Number of observations = 132. Log likelihood = − 195.91. Likelihood Ratio Chi-squared with 21 degrees of freedom = 97.59, with the probability of a greater chi-squared = 0.0000. Pseudo $R^2$ = 0.199. Tobin's $\sigma$, the ancillary parameter for the Tobit model, is estimated to be 27.25 with standard error equal to 3.24. There were 95 left-censored observations at PROD3RD = 0, and there were 37 uncensored observations where PROD3RD > 0.

Reestimating the models of Tables 6.2 and 6.3, dropping the term for the square of IMPSC, we can ask if the effect of import competition *on the underlying Tobit index* is simply a linear one for PROC3RD and PROD3RD. Even with the simpler model, import competition does not have a significant effect on PROC3RD. The model for PROD3RD shows a linear negative effect for IMPSC on the index for Title III process R&D as seen in Table 6.5. The simulated results for the unconditional expected value of the R&D are depicted in Figure 6.11.

Large US corporations may be expressing their discomfort with such import competition when they voice concerns — in the context of environmental regulation — about the competitive pressures from imports. Such pressures are especially worrisome to the companies when US regulations require emissions standards that the foreign competitors do not face. On the whole, the estimations show that greater import competition is associated with less emissions-reducing R&D investment, *ceteris paribus*. To the extent that it is unprofitable to compete with the foreign firms and invest in R&D for improved environmental performance, the results lend support to industry's position. Jaffe *et al.* (1995) conclude that the competitiveness of firms does not appear to have suffered greatly because of environmental regulation. Our result suggests that there may well be a cost associated with maintaining that competitiveness. Namely, it appears that firms facing import competition have cut their environmental R&D.

The significant patterns for the coefficients for the measures of Title III emissions problems, and for seller concentration and import competition, imply that the problems and competition as measured by seller concentration and the pressure of import competition affect environmental R&D effort, even after the control for broad industry effects. No simple generalization can be made about

*Figure 6.11. Unconditional Expected Value of PROD3RD as a Function of IMPSC Using Specification of Table 6.5*

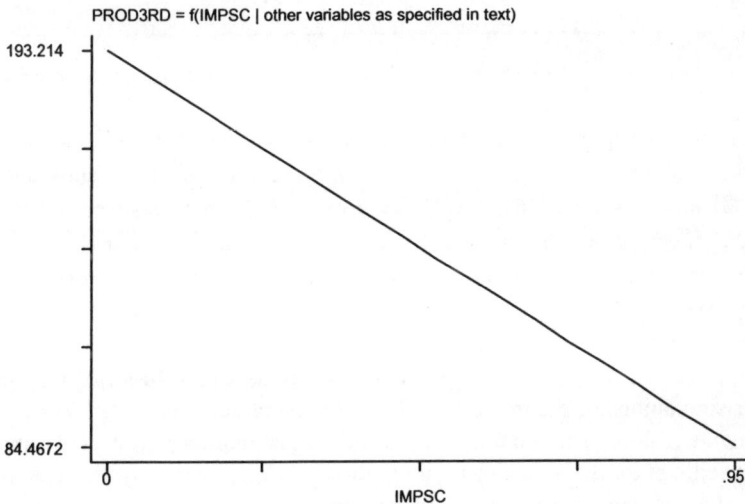

PROD3RD = f(IMPSC | other variables as specified in text)

'Schumpeterian competition' that occurs in concentrated industries as contrasted with the more traditional competition in industries with relatively low seller concentration. Other things held constant, there is evidence of substantial Title III research investments given both high and low seller concentration for some types of R&D, with the least R&D for intermediate levels of seller concentration. However, for all types of R&D, high seller concentration is associated with more R&D than what would occur with lower concentration. In our model, the reason would be that emissions problems and dynamic competition from large R&D intensive firms with market power increase the probability that a company will not, *ceteris paribus*, have sufficient technology to match the regulatory requirements given the state of the art processes and products. Such a company then invests more than it would otherwise invest to improve its chances of meeting regulatory requirements.

*Table 6.6. Predicted Title III R&D in Millions of 1992 Dollars for a Company with Substantial Title III Emissions\**

| Type of Title III R&D | Conditional Expected Value Given RIII > 0 | Unconditional Expected Value |
|---|---|---|
| Background Research | 2.63 | 0.207 |
| Process R&D | 12.6 | 12.5 |
| Product R&D | 171.4 | 171.4 |

\*For a company with the variables set as in the simulations of the models in Tables 6.1, 6.2, and 6.3.

In subsequent chapters, we shall pursue the implication of the estimations in Tables 6.1 through 6.3 and the extensions of the estimates in the subsequent tables. Given the estimations, a role for government, when society wants to reduce the negative externalities caused by hazardous air emissions, is to set standards that cause firms to accept the uncertainty of R&D investment in reducing toxic emissions. The evidence suggests that once they accept that uncertainty, their investments in environmental R&D are fairly substantial.

To illustrate that fact, consider again our three models in Tables 6.1, 6.2, and 6.3. Our simulations of the models are for the unconditional expected values of environmental R&D. Table 6.6 below shows the unconditional expected values using the settings for the variables used in our simulations, with any variable that changed in a simulation above now held constant at its mean, excepting NTAPC that is set, as before, at a standard deviation above its mean. For comparison, Table 6.6 also shows the conditional expected values — conditional on the presence of environmental R&D of the type examined — for the three types of

Title III R&D. The conditional and unconditional values are fairly close because firms with substantial Title III emissions problems are likely to do Title III R&D.

*Table 6.7. Predicted Title III R&D in Millions of 1992 Dollars for a Company with Substantial Title III Emissions and with the Other Types of Title III R&D\**

| Type of Title III R&D | Conditional Expected Value Given RIII > 0 | Unconditional Expected Value |
|---|---|---|
| Background Research | 5.49 | 3.52 |
| Process R&D | 22.1 | 22.1 |
| Product R&D | 204.8 | 204.8 |

\*For a company with the variables set as in the simulations of the models in Tables 6.1, 6.2, and 6.3, except that the dummy variables indicating the other types of Title III R&D are now set equal to 1.

*Figure 6.12. Unconditional Expected Value of BACK3RD (Including Hazard Rate's Effect) as a Function of NTAPC*

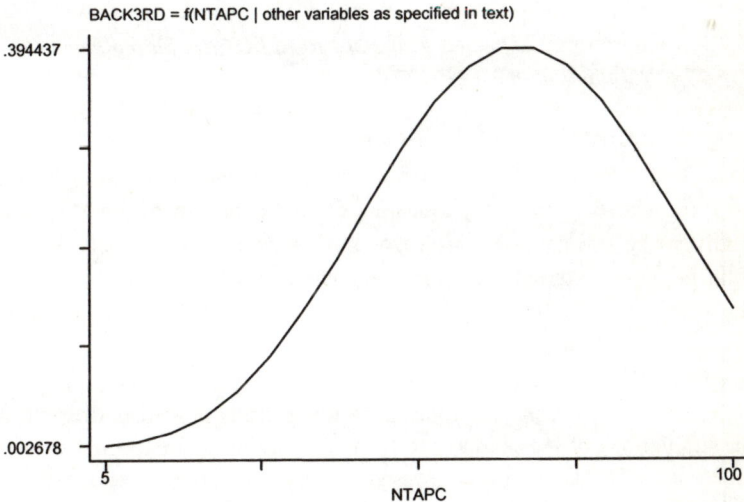

BACK3RD = f(NTAPC | other variables as specified in text)

Note that for companies doing two or three of the types of Title III R&D, the expected value of any particular type of Title III R&D will be considerably greater, because in our foregoing simulations we set the dummy variables indicating other types of Title III R&D to zero. To show the comparison, Table 6.7 provides the predictions for each particular type of Title III R&D when the company does the other two types of Title III R&D. The predictions in Tables 6.6 and 6.7 *are* for very special firms. They do substantial amounts of environmental R&D, and are a special group of firms. In particular, in our sample of manufacturing firms, companies with sales, seller concentration, and import competition at or greater than the averages for the sample, and with Title III emissions a standard deviation or more above the average, have total R&D expenditures that average $660 million dollars and sales that average $15,400 million dollars.

To this point the predictions have been presented under the assumption that our Tobit models are reasonably complete models. If instead there are explanatory variables that are important and that have been omitted from the model, then it is entirely possible that the effects of the hazard rate are not simply error components that result from selection bias. Possibly the effects of the hazard rate are systematic effects of left-out explanatory variables that belong in the model. In that case, the predictions of RIII should include the hazard rate effect; the simulations would be done with the hazard rate set at its mean in the sample. The results would be somewhat larger predictions for the underlying background research (where the hazard effect is positive) and considerably smaller predictions for process and product R&D (where the hazard effect is negative).

*Figure 6.13. Unconditional Expected Value of PROC3RD (Including Hazard Rate's Effect) as a Function of NTAPC*

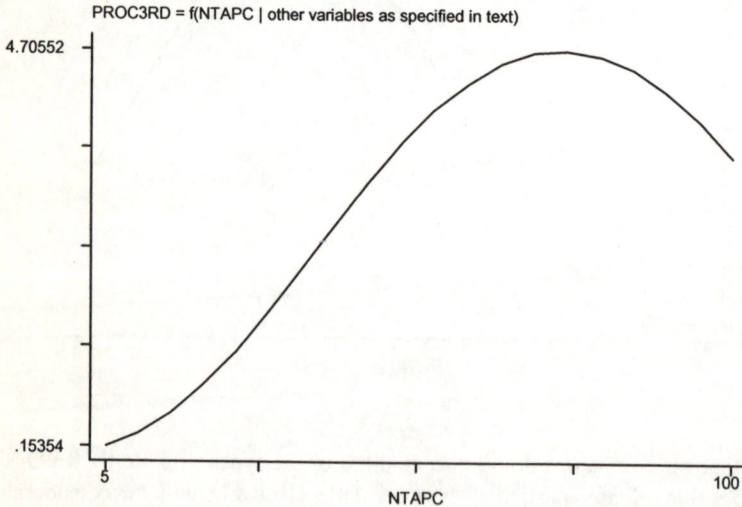

PROC3RD = f(NTAPC | other variables as specified in text)

The following figures and tables replicate the simulations in this chapter with the exception that the hazard rate's effect (with the hazard rate set at its mean for the sample) is added back into the prediction. The comparison of the two sets of simulations shows that prediction of the amount of Title III R&D to be expected in specified circumstances depends on whether we believe the hazard effect reflects selection bias or instead the effect of left out variables in the Tobit model. The signs and significance of the partial effects of our explanatory variables are quite robust to a variety of specifications and models from OLS to Tobit or to the generalization of Tobit using the 'intreg' procedure with robust standard errors. A few details about the nonlinearity of the results aside, we are therefore reasonably confident that greater emissions, greater sales, greater concentration, greater import competition and so on will have the direction and significance of effects described in the estimations. The predicted amount of Title III R&D in specified circumstances — as contrasted to the sign and significance of the partial effect on that prediction caused by changing a variable — is sensitive to our interpretation of the effect of the hazard rate. The sensitivity is shown by the comparison of Figures 6.12 through 6.22 and Tables 6.8 and 6.9 with Figures 6.1 through 6.11 and Tables 6.6 and 6.7.

*Figure 6.14. Unconditional Expected Value of PROD3RD (Including Hazard Rate's Effect) as a Function of NTAPC*

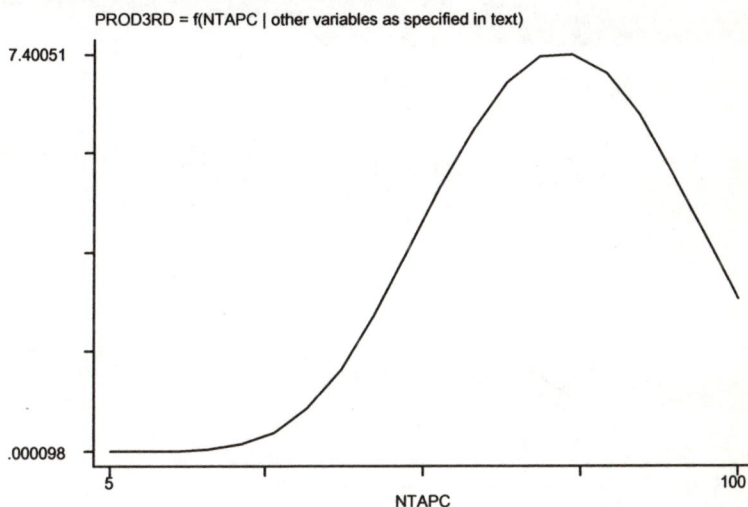

PROD3RD = f(NTAPC | other variables as specified in text)

*Figure 6.15. Unconditional Expected Value of BACK3RD (Including Hazard Rate's Effect) as a Function of SALES*

BACK3RD = f(SALES | other variables as specified in text)

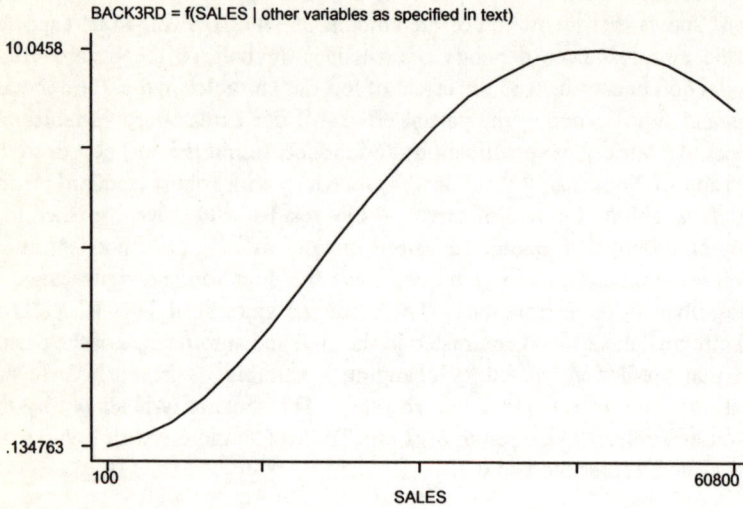

10.0458

.134763

100                                                              60800

SALES

*Figure 6.16. Unconditional Expected Value of PROC3RD (Including Hazard Rate's Effect) as a Function of SALES*

PROC3RD = f(SALES | other variables as specified in text)

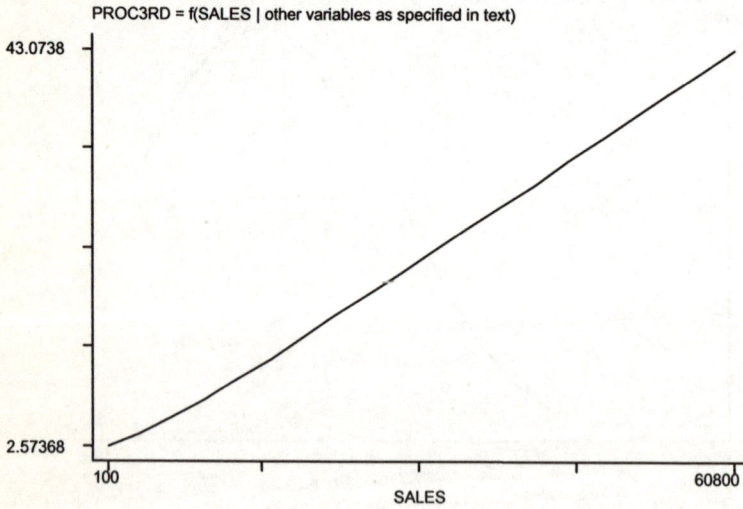

43.0738

2.57368

100                                                              60800

SALES

*Figure 6.17. Unconditional Expected Value of PROD3RD (Including Hazard Rate's Effect) as a Function of SALES*

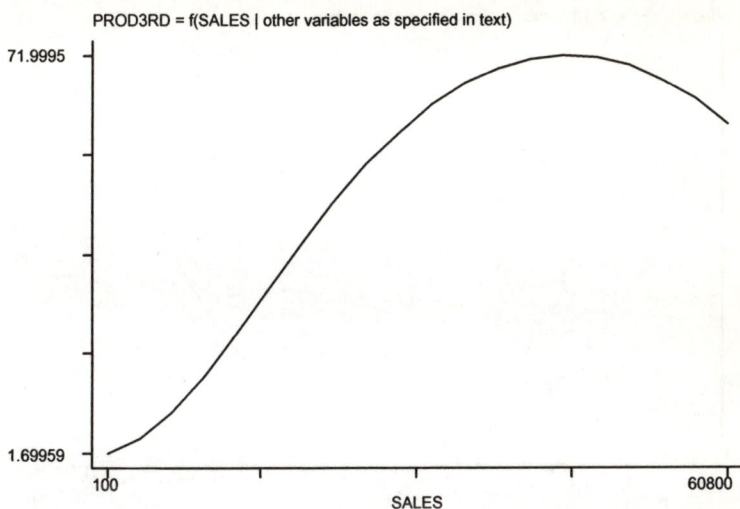

PROD3RD = f(SALES | other variables as specified in text)

71.9995 –

1.69959 –

100                                 SALES                                 60800

*Figure 6.18. Unconditional Expected Value of BACK3RD (Including Hazard Rate's Effect) as a Function of CR4C*

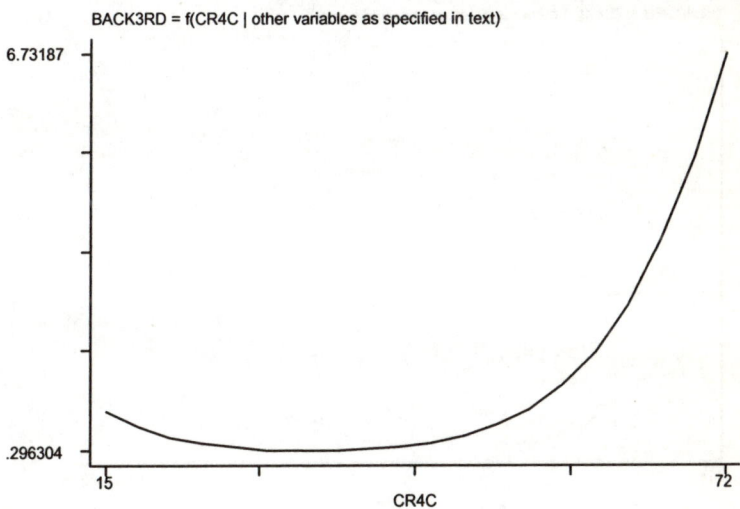

BACK3RD = f(CR4C | other variables as specified in text)

6.73187 –

.296304 –

15                                 CR4C                                 72

*Figure 6.19.   Unconditional Expected Value of PROC3RD (Including Hazard Rate's Effect) as a Function of CR4C*

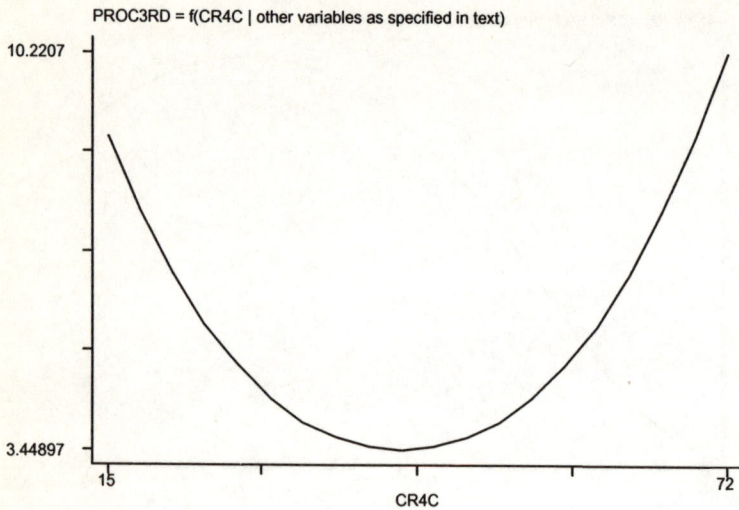

PROC3RD = f(CR4C | other variables as specified in text)

*Figure 6.20.   Unconditional Expected Value of PROD3RD (Including Hazard Rate's Effect) as a Function of CR4C*

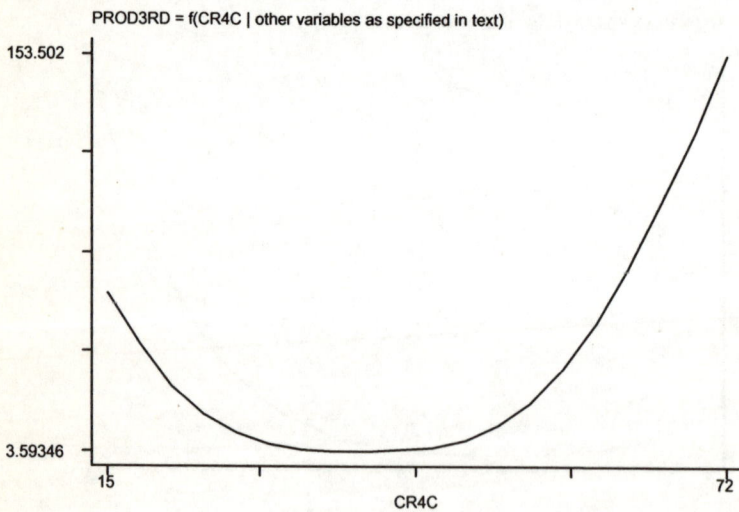

PROD3RD = f(CR4C | other variables as specified in text)

*Figure 6.21.  Unconditional Expected Value of BACK3RD (Including Hazard Rate's Effect) as a Function of IMPSC*

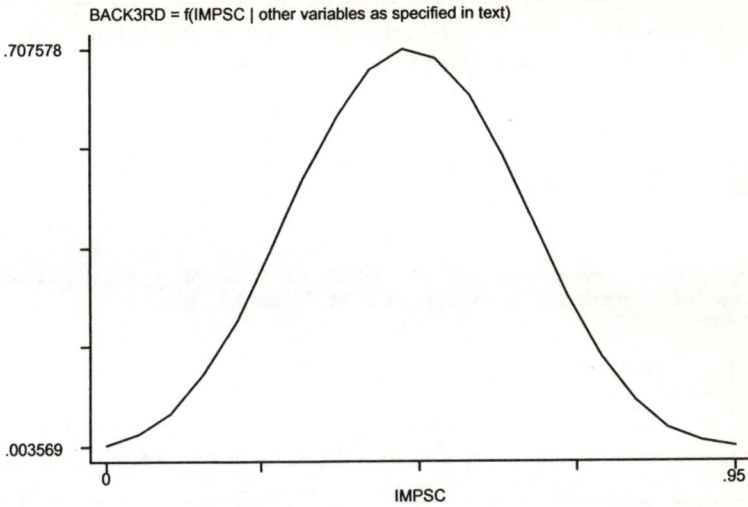

BACK3RD = f(IMPSC | other variables as specified in text)

*Figure 6.22.  Unconditional Expected Value of PROD3RD (Including Hazard Rate's Effect) as a Function of IMPSC Using Specification of Table 6.5*

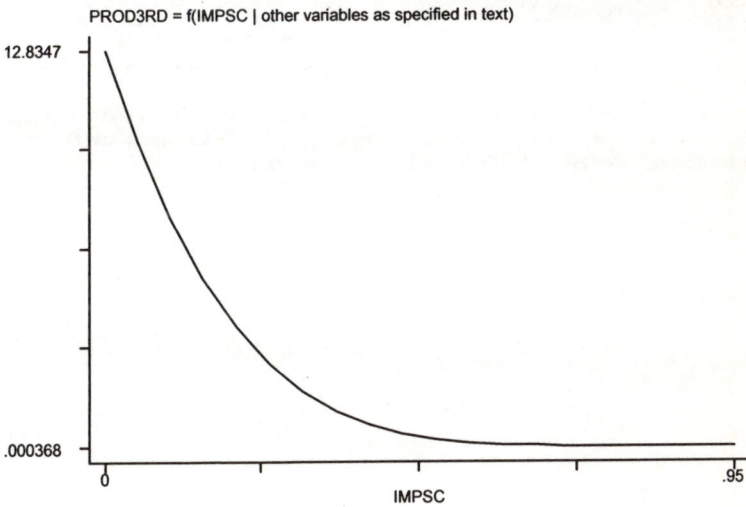

PROD3RD = f(IMPSC | other variables as specified in text)

*Table 6.8.  Predicted Title III R&D (Including Hazard Rate's Effect) in Millions of 1992 Dollars for a Company\**

| Type of Title III R&D | Conditional Expected Value Given RIII > 0 | Unconditional Expected Value |
|---|---|---|
| Background Research | 2.79 | 0.298 |
| Process R&D | 5.38 | 3.66 |
| Product R&D | 15.9 | 3.77 |

*For a company with the average values for NTAPC, SALES, CR4C, and IMPSC, and with all other variables set as in the simulations of the models in Tables 6.1, 6.2, and 6.3.

*Table 6.9.  Predicted Title III R&D (Including Hazard Rate's Effect) in Millions of 1992 Dollars for a Company with the Other Types of Title III R&D\**

| Type of Title III R&D | Conditional Expected Value Given RIII > 0 | Unconditional Expected Value |
|---|---|---|
| Background Research | 5.96 | 4.19 |
| Process R&D | 12.3 | 12.2 |
| Product R&D | 27.6 | 19.2 |

*For a company with the average values for NTAPC, SALES, CR4C, and IMPSC, and with all other variables set as in the simulations of the models in Tables 6.1, 6.2, and 6.3 except that the dummy variables indicating the other types of Title III R&D are now set equal to 1.

# 7.  Cooperative R&D

The theory of Chapter 5 and the tests of the theory in Chapter 6 point to dynamic competition as a spur to environmental R&D.  Although a priori theory and some evidence support the possibility that dynamic competition could be most intense in less concentrated markets, on the whole, the tests support Schumpeter's hypothesis that dynamic competition comes from firms in more concentrated markets.  In this chapter, we shall ask if cooperative ventures to work on environmental R&D have lessened that desirable competition, or instead have allowed the R&D to be conducted more efficiently.  Chapter 4 described the extent to which industrial firms in our sample have joined forces in cooperative ventures to reduce industrial emissions.  This chapter adds to that description and asks if public policy expressing concern about the emissions brought about such cooperation and if the cooperation improves the performance of environmental R&D or instead simply reduces desirable competition.[1]

## COOPERATION AND THE LAW

A research joint venture arguably improves economic performance.  Such a venture may bring forth socially desirable new products and processes more efficiently than individual firms investing noncooperatively.  Research joint ventures may reduce risk and uncertainty, achieve economies of scale and scope, allow companies to share costs and make very costly R&D possible, eliminate wasteful duplication of research efforts, and improve the appropriation of returns from innovation.

Such socially desirable improvements in innovative investments were among the reasons given in support of new laws in the United States (Scott, 1993, pp.

---

[1] Scott (1996) addresses the question with the data initially available — the data for the 55 firms reporting Title III process R&D.  In this chapter, we can address the question anew with all of the data in hand — reports from the 58 firms reporting process R&D to reduce emissions (of which the 55 Title III process R&D firms are a subset), as well as reports from companies doing background research and doing product R&D to reduce emissions.

187–202). The National Cooperative Research Act (NCRA) of 1984 (US 98th Congress, 1984) and its amendment, the National Cooperative Research and Production Act (NCRPA) of 1993 (US 103rd Congress, 1993) were designed to encourage the formation of innovative joint ventures by reducing their potential antitrust liabilities.[2] However, the research ventures could also eliminate desirable competition that spurs innovative investment (Scott, 1993). In this chapter, I explore the reasons that firms form research joint ventures by focusing on a particular type of venture — namely research joint ventures that study emissions or develop new processes or products to reduce toxic air emissions.

We shall begin by comparing NCRA/NCRPA research joint ventures before and after the Clean Air Act Amendments (CAAA) of 1990 (U.S. 101st Congress, 1990). The comparison shows the effect of the CAAA on the formation of research joint ventures, and thereby gives some insight into the power of government to stimulate joint research with the goal of reducing negative externalities. We shall then model the determinants of the probability of joining a cooperative venture. Do the factors include those anticipated by theory and policy — factors such as the riskiness of R&D, difficulties appropriating its returns, the costliness of R&D, and the importance of scale economies in research and production? The evidence is then used to address the question of whether on balance research joint ventures exploring emissions problems are likely to improve economic performance.

## THE CLEAN AIR ACT AND RESEARCH JOINT VENTURES

In earlier research (Scott, 1988), I studied the first 61 research joint ventures that were filed under the NCRA. Those filings covered the 18 months from January 1985 through June 1986. Thus, the ventures were formed well before the CAAA identified a list of hazardous air pollutants emitted by industry that with the passage of the Act became the target of public scrutiny for future regulation. As explained in Chapter 2, the CAAA of 1990 listed the targeted set of hazardous air pollutants in Title III and charged EPA with developing emissions standards. In the survey, R&D directors reported their R&D activity related to Title III chemicals was in anticipation of regulation in order to avoid future problems. For this look at the NCRA ventures, I shall focus on just these hazardous chemicals that have come under increased regulatory scrutiny — the Title III hazardous air pollutants. The list of Title III pollutants is provided in the appendix to Chapter

---

[2] Ventures filing under the Acts with the US Federal Trade Commission and the US Department of Justice are, if challenged under the US antitrust laws, to be judged under a Rule of Reason (rather than per se) standard and would be liable for just single (rather than treble) damages if found to be in violation of the law.

3.  As discussed in Chapters 2 and 3, Title III lists 190 chemicals (for example, benzene) or groups of chemicals (for example, arsenic compounds).

For an initial look at the impact of government regulation on the formation of research joint ventures, I shall compare the joint ventures in the first 18 months of reported ventures with the ventures filed under the NCRA in the 18 month period from February 15, 1991 to August 15, 1992.[3] I shall look for ventures related to the Title III hazardous air pollutants. The second period begins one quarter (three months) after the passage of the CAAA on November 15, 1990; the lag of one quarter allows time for formation and filing by ventures responding to the US Congress's list of Title III pollutants. The Title III toxic chemicals have been pervasive in US industry for many years. Comparing the environmental research joint ventures of the two 18 month periods provides one measure of the impact of Congress's clear 1990 statement in the CAAA that these hazardous air pollutants would henceforth be subject to scrutiny and to evolving regulations.

Of the 61 ventures filed during the first 18 month period, 20 were related to environmental issues, using the criterion that, in my opinion, the ventures were explicitly studying processes or products in order to understand and ultimately reduce environmental problems such as air pollution from toxic emissions. Of those 20 ventures, 16 were concerned, at least in part, with hazardous air emissions. Other environmental issues included land and water pollution, and metal and plastics recycling. Just one of the 61 ventures was clearly related to any of the Title III chemicals. That is, with the exception of the one venture which focused on benzene emissions from late model passenger cars, Title III chemicals were never specifically mentioned as the focus of research, whether to gather information about the chemical or to develop new processes or products to reduce its emission. Further, benzene is a chemical that is mentioned not only in Title III, but is regulated by 'Title II — Provisions Relating to Mobile Sources' (see Section 219, CAAA, 1990, p. 2493, for example) and has been a well-known chemical in automobile exhaust emissions that were covered in earlier versions of the Clean Air Act. Surely, some of the other ventures among the 61 were concerned in part with some of what would come, about six years later, to be known as the Title III chemicals although the chemicals were not mentioned by name; ventures that seek to understand and reduce toxic air emissions as a general goal would probably consider chemicals that were not yet regulated but known to be toxic. But when one compares the ventures of the first period with those in the sample period after the CAAA, one perceives a noticeable increase in activity focused on Title III toxins, suggesting the power of the government to stimulate research about negative externalities. The evidence implies that the

---

[3] My sources for information about the ventures were US General Services Administration (1985, 1986, 1991, 1992) and US Department of Commerce (1993).

government can channel resources indirectly through the rational responses of firms to the government's regulations.

I must emphasize that the foregoing assessment for the first period as well as the following one for the second period is a rough assessment for several reasons in addition to the possibility that Title III chemicals were involved even when not mentioned specifically. First, one could debate about even the proper number of ventures, because in some cases several of the ventures were filed by the same organization. For the most prominent example, the Motor Vehicle Manufacturers Association of the United States (MVMA) filed 15 separate and quite distinct ventures during the first period — all on February 8, 1985. At least 11 of the 15 ventures were related to environmental issues — mostly focused on exhaust emissions. Thus, the proportion of ventures taken by environmental projects would be considerably less if the ventures were counted totaling the number of primary research organizations (such as the MVMA or Bellcore) instead of the number of research projects that were filed under the NCRA as ventures, some of which have the same primary research organization and in some cases even exactly the same members. Second, some of the ventures that are clearly related to environmental concerns are just as clearly only tangentially related. For example, the Portland Cement Association lists numerous objectives generally having to do with improved production methods. One among the several objectives is improving compliance with environmental regulations. Third, several ventures simply state the goal of improving production methods but make no mention of improving environmental performance. Possibly environmental concerns would be an issue for these ventures, but they are not counted among the ventures with environmental concerns.

Ninety-six ventures were filed under the NCRA with the Department of Justice and the Federal Trade Commission from February 15, 1991 to August 15, 1992 (with publication dates in the *Federal Register* from March 12, 1991 to October 6, 1992). Of the 96 ventures filed during this second 18 month period, 35 were obviously directly related to environmental issues, and 27 of those were, at least in part, concerned with hazardous air emissions. In marked contrast to the first period, 10 were clearly related to the Title III chemicals because the chemicals were mentioned by name. Another 12 were probably, or in some cases almost certainly, concerned with Title III chemicals. Although they were not mentioned by name in those 12 cases, Title III chemicals were among the general class of toxic air pollutants that the ventures were researching. Among the ventures not counted in the 35 environmental ventures were some that surely could be construed as indirectly addressing environmental concerns. For example, in a venture with the University of California at Berkeley, 14 organizations — mostly shipping or ship-building companies — conducted R&D to improve design and repair of ships' structural components that are affected by corrosion and fatigue. Arguably such R&D could reduce environmental damage from shipping spills, although I have not counted the venture as one related to environmental concerns. For another example of a venture not counted among the environmental ventures, the Advanced Reactor Corporation constituted a venture of 16 companies formed to conduct R&D for a new generation of nuclear power plants. Again, one could argue that the new designs would reduce at least the potential for environmental damage.

The survey of the ventures formed during the two periods suggests two factual generalizations. First, using a very conservative counting method, roughly a third of the NCRA research joint ventures address environmental concerns, and further, one-third is clearly a lower bound for the proportion of environmental ventures. Link (1996) shows somewhat lower percentages for environmental ventures. Even when the word 'environmental' did not appear in the *Federal Register* listing, I have counted ventures studying the emissions of a chemical when that chemical was listed by the government as one that companies would have to start cleaning up. Also, I have included ventures working on issues related to environmental concerns more generally, such as a venture exploring possibilities for recycling plastics. For a second factual generalization, the proportion of environmental ventures that addressed concerns about Title III hazardous air pollutants jumped dramatically after the passage of the CAAA of 1990. Thus, a third generalization that interprets the facts is that the legislative expression of government concern and intent to develop regulations for toxic chemicals prompts firms to join forces in joint ventures that address the problems posed by the chemicals.

## THE PROBABILITY OF JOINING A COOPERATIVE RESEARCH VENTURE

I shall explore the reasons that manufacturers join cooperative ventures with other companies by using a probit model to describe the relations between the probability of joining a cooperative venture to reduce emissions and the various characteristics of companies, their environmental R&D projects, and their industries.

We expect that companies join ventures when it is otherwise difficult to appropriate the returns from innovation, when research is especially costly, when it is especially risky, and when there are marked efficiencies to be gained by increasing the size of the research effort. Important variables would, a priori, include difficult appropriability conditions (*APP*), the costliness of the R&D projects (*COST*), the riskiness of the projects (*RISK*), and the importance of the size of the research effort for realizing efficiencies (*EFF*). Further, the availability of knowledge from outside — i.e., from knowledge spillovers from other firms — about the firm's R&D problems might affect the company's willingness to join research ventures, and we shall explore the importance of control for the likelihood of such knowledge spillovers  using the variable *SPILLC* which is defined in the Glossary.  Important industry characteristics might include variables that we have used in earlier chapters to capture the extent of structural competition in the domestic market (*CR4C*) and the extent of import competition (*IMPSC*).  Important company characteristics might include the size of the company's R&D effort (*RD*) and the size of the company (*SALES*) as we have defined them earlier.

We shall use the survey data described in Chapters 3 and 4 to test the hypotheses about the determinants of the probability of joining in cooperative environmental R&D.  For our hypothesis tests, we define the following new variables.

CoopBack equals 1 if a respondent chose the first answer to survey question #5 — i.e., if the company said that it had background emissions research that was performed in a cooperative venture with other firms.  If a company had background emissions research but did not perform any of that research in cooperation with other firms, then CoopBack equals 0.

CoopProc equals 1 if a respondent chose the first answer to survey question #18 — i.e., if the company said that it had process R&D to reduce emissions that was performed in a cooperative venture with other firms.  If a company had process R&D to reduce emissions but did not perform any of that R&D in cooperation with other firms, then CoopProc equals 0.

CoopProd equals 1 if a respondent chose the first answer to survey question #31 — i.e., if the company said that it had product R&D to reduce emissions that was performed in a cooperative venture with other firms.  If a company had product R&D to reduce emissions but did not perform any of that R&D in cooperation with other firms, then CoopProd equals 0.

RiskBack, RiskProc, and RiskProd are defined respectively for companies reporting background emissions research, process R&D to reduce emissions, and product R&D to reduce emissions.  For companies with the particular type of research, the RISK variable equals 1 if the company reported that the work was more risky in comparison with the riskiness of its other research or R&D projects. Otherwise the variable equals 0.  Thus, the variable is defined as 1 for those companies choosing the third response to questions #9, #22, or #35, and as 0 for those companies choosing the first or second response to those questions.

AppBack, AppProc, and AppProd are defined respectively for companies reporting background emissions research, process R&D to reduce emissions, and product R&D to reduce emissions.  For companies with the particular type of research, the APP variable equals 1 if the company reported that it was difficult or very difficult to realize a normal return from its investments in the research or R&D.  Otherwise the variable equals 0.  Thus, the variable is defined as 1 for those companies choosing the third or the fourth response to questions #6, #19, or

#32, and as 0 for those companies choosing the first or second response to those questions.

CostBack, CostProc, and CostProd are defined respectively for companies reporting background emissions research, process R&D to reduce emissions, and product R&D to reduce emissions. For companies with the particular type of research, the COST variable equals 1 if the company reported that as compared with its other research or R&D projects the emissions work was more costly. Otherwise the variable equals 0. Thus, the variable is defined as 1 for those companies choosing the third response to questions #8, #21, or #34, and as 0 for those companies choosing the first or second response to those questions.

EffBack, EffProc, and EffProd are defined respectively for companies reporting background emissions research, process R&D to reduce emissions, and product R&D to reduce emissions. For companies with the particular type of research, the EFF variable equals 1 if the company reported that as compared with its other research or R&D projects efficiencies from the size of the research effort were more important for the emissions work. Otherwise the variable equals 0. Thus, the variable is defined as 1 for those companies choosing the third response to questions #10, #23, or #36, and as 0 for those companies choosing the first or second response to those questions.

To control for industry effects in the small samples for which we have the foregoing data about cooperative activity, risk, appropriability, costliness of research, and efficiencies of size, we define three dummy variables — LOW, MODERATE, and HIGH — that place each of the two-digit SIC industries into a low, moderate, or high R&D intensity group. All 132 respondents with manufacturing activity were grouped by primary two-digit industry, and then the average of R&D/SALES was computed for the companies within each two-digit manufacturing group and for those firms in our sample with manufacturing activity but with their primary industry in nonmanufacturing. The low intensity group included industries with an average R&D intensity ranging from 0.0046 through 0.011 — roughly 1 percent of sales or less. Those industries included SIC 20, 25, 26, 29, 30, and 33. The moderate intensity group included industries with an average R&D intensity ranging from 0.022 through 0.027 — between 2 and 3 percent of sales. The moderate intensity group included SIC 22, 24, 34, and the nonmanufacturing group. The high intensity group included industries with an average R&D intensity in our sample that ranged from 0.030 through 0.095 — from 3 to 9.5 percent of sales. The high intensity group for our sample of firms included SIC 28, 35, 36, 37, 38, and 39. One might hypothesize that companies in industries in the low or moderate intensity groups would be more likely to join with other firms in cooperative environmental research because their own research resources were relatively small.

Estimating the probit models for the probability of cooperative activity in the various types of environmental research provides support for some of the hypotheses advanced to explain cooperative activity. For this exploratory look at the determinants of the probability of a company joining a cooperative venture doing environmental R&D, the specifications reported in Tables 7.1 through 7.4 are for the most part parsimonious ones that include just variables that proved significant in larger models. Thus, for each type of research, a model including all the various variables for SALES, RD, APP, RISK, COST, EFF, SPILLC, industry effects, response bias, and industry characteristics such as seller concentration and import competition were examined. Variables that were not significant were dropped, leaving the specifications that are reported.

*Table 7.1. The Probit Model of CoopBack*

| Variable | Estimated Coefficient | Asymptotic t-ratio (z) | Probability > \|z\| |
|---|---|---|---|
| SALES | 0.000171 | 1.63 | 0.104 |
| RD | 0.00519 | 1.43 | 0.153 |
| AppBack | 4.37 | 1.68 | 0.094 |
| RiskBack | 2.45 | 2.01 | 0.044 |
| LOW | 2.11 | 1.48 | 0.139 |
| MODERATE | 4.76 | 2.00 | 0.045 |
| Hazard Rate | 3.67 | 1.50 | 0.135 |
| CR4C | – 0.0688 | – 1.36 | 0.173 |
| Constant | – 10.1 | – 1.81 | 0.071 |

Number of observations = 41
Log likelihood = – 8.657
Likelihood Ratio Chi-squared statistic with 8 degrees of freedom = 30.37 (probability of a greater chi-squared = 0.0002)
Pseudo $R^2$ = 0.637

Highlighting findings in the tables, we have the following results, other things being the same. Because larger firms typically have more R&D activities, in these probit models size could be seen as a simple control for a greater likelihood that some cooperative activity would appear among the greater number of activities. Also, if larger firms have more assets to offer to research partnerships than smaller firms, they may be more likely to be included in RJVs. In our sample, larger firms as measured by both sales and total company-financed R&D are more likely to cooperate in background emissions research. For process R&D, company size by sales is not important, but size as measured by R&D expenditures is associated with more cooperative process R&D to reduce emissions. For emissions-reducing product R&D, company size is again important, but companies of intermediate size do more cooperative R&D than the largest and the smallest firms in our sample.

*Table 7.2. A More Parsimonious Probit Model of CoopBack*

| Variable | Estimated Coefficient | Asymptotic t-ratio (z) | Probability > \|z\| |
|---|---|---|---|
| SALES | 0.000200 | 2.53 | 0.011 |
| AppBack | 4.026 | 1.80 | 0.071 |
| RiskBack | 1.34 | 1.60 | 0.110 |
| LOW | 1.085 | 1.11 | 0.267 |
| MODERATE | 4.173 | 2.10 | 0.036 |
| Hazard Rate | 2.814 | 1.59 | 0.111 |
| Constant | − 9.865 | − 2.21 | 0.027 |

Number of observations = 41
Log likelihood = − 10.34
Likelihood Ratio Chi-squared statistic with 6 degrees of freedom = 27.01 (probability of a greater chi-squared = 0.0001)
Pseudo $R^2$ = 0.5665

Firms reporting difficulties appropriating the returns from their environmental R&D are more likely to join cooperative ventures doing background research on emissions or process R&D to reduce emissions, but in our sample appropriability difficulties are not important in the decision to do cooperative product R&D.

Companies reporting that their background research on emissions is very risky are more likely to join cooperative ventures for that research. Similarly, companies reporting that their emissions-reducing product R&D is very risky are more likely to join cooperative ventures for that R&D. But, in our sample, the riskiness of the emissions-reducing process R&D does not appear an important determinant of the probability of cooperative R&D.

The costliness of R&D and the importance of efficiencies from the size of the research effort are not important in our sample for the probability of joining cooperative ventures for background research or for process R&D. Surprisingly, for product R&D, costliness and important efficiencies from size are negatively related to the decision to join a venture. One suspects the result has something to do with the peculiarities of product R&D aimed at emissions reduction. In Chapter 6, we have seen that the product R&D is especially sensitive to seller concentration, being greatest for very high concentration. The R&D doing firms in such industries may achieve the necessary size for successful operation without sharing knowledge and cooperating with their Schumpeterian rivals.

*Table 7.3. A Probit Model of CoopProc*

| Variable | Estimated Coefficient | Asymptotic t-ratio (z) | Probability > \|z\| |
|----------|----------------------|------------------------|---------------------|
| RD | 0.00360 | 2.55 | 0.011 |
| AppProc | 1.75 | 2.04 | 0.041 |
| MODERATE | 1.55 | 1.90 | 0.057 |
| SPILLC | − 0.00955 | − 1.50 | 0.134 |
| Constant | − 1.34 | − 1.13 | 0.259 |

Number of observations = 58
Log likelihood = − 13.78
Likelihood Ratio Chi-squared statistic with 4 degrees of freedom = 22.50 (probability of a greater chi-squared = 0.0002)
Pseudo $R^2$ = 0.4495

As expected, companies with primary activities in industries in the low to moderate R&D intensity categories are more likely to join cooperative ventures than those in the high intensity category. The effect for the moderate intensity category is important for all three types of environmental R&D, while the effect for the low intensity category appears only with background research.

Following reasoning directly analogous to Chapter 6's discussion of the estimation of the Tobit model with selection into the sample, the appropriate hazard rate from the response model is entered into the probit model to control for response bias. The hazard rate appears somewhat significant in the model of the probability of joining cooperative activity in background research, but it does not appear to be important at all in the other models. The findings for the explanatory variables are essentially unchanged with or without control for response and when the probability of response rather than the hazard rate is used for the response bias control. Further, StataCorp (2001, Volume 2, pp. 580–594) has the option of estimating robust standard errors with the probit model. For the probit models presented here, estimating the robust standard errors does not change the story from what the simple probit models show, except that the significance of the coefficients increases.

Seller concentration is never really important, although it enters with a negative sign and a z-statistic greater than one in absolute value when added to the specification background research. Import competition also does not appear to be important for the probability of joining a cooperative research venture. Similarly, our exploratory measure of the extent of knowledge spillovers is never very important. In the specification for process R&D it has a negative sign and a z-statistic exceeding one in absolute value.

*Table 7.4. A Probit Model of CoopProd*

| Variable | Estimated Coefficient | Asymptotic t-ratio (z) | Probability > |z| |
|---|---|---|---|
| SALES | 0.000400 | 2.29 | 0.022 |
| SALES $^2$ | $-5.70 \times 10^{-9}$ | $-2.26$ | 0.024 |
| RiskProd | 3.66 | 2.36 | 0.018 |
| CostProd | $-1.75$ | $-1.79$ | 0.074 |
| EffProd | $-4.39$ | $-1.68$ | 0.093 |
| MODERATE | 4.36 | 2.37 | 0.018 |
| Constant | $-3.59$ | $-2.60$ | 0.009 |

Number of observations = 39
Log likelihood = $-8.539$
Likelihood Ratio Chi-squared statistic with 6 degrees of freedom = 22.50 (probability of a greater chi-squared = 0.0010)
Pseudo $R^2$ = 0.5685

To augment the findings in the preceding tables, we can look back to the three models in Tables 6.1, 6.2, and 6.3 of Chapter 6. Using those models we can now predict the three types of Title III R&D — BACK3RD, PROC3RD, and PROD3RD — for each of the firms that *have* each particular type of Title III R&D. We can then measure the difference DIFF between the actual amount of each type of R&D from its predicted amount (the conditional expectation given RIII > 0) for those firms having each type of R&D. Then, for just those firms having the particular type of Title III R&D, regressing DIFFBACK on CoopBack, DIFFPROC on CoopProc, and DIFFPROD on CoopProd, we can ask if the presence of cooperative R&D activity increases environmental R&D significantly. It clearly does for background research and process R&D as seen in Table 7.5.

Expected background research for a company is about $4 million dollars more in 1992 dollars when a company does cooperative background research. Expected emissions-reducing process R&D for a company is $3.9 million more given that the company is involved in cooperative process work. These estimated increases are substantial relative to the expected magnitudes of Title III R&D predicted at the end of Chapter 6 for a company with the characteristics specified there. The analogous specification for product R&D (not shown) does have a positive sign for cooperative activity, but the effect is thoroughly insignificant. Emissions-related product R&D does not appear to be significantly related to cooperative activity, perhaps reflecting the unexpected negative effects on cooperative product R&D of research costliness and the importance of efficiencies of size.

*Table 7.5. OLS Model for Expected Title III R&D Difference and Cooperative Activity*

---

Dependent Variable DIFFBACK

---

| Variable | Coefficient | t-statistic with 39 degrees of freedom | Probability > \|t\| |
|---|---|---|---|
| CoopBack | 3.98 | 2.23 | 0.031 |
| Intercept | − 2.98 | − 3.23 | 0.003 |

---

$R^2 = 0.113$;  Adjusted $R^2 = 0.0905$
F for the equation as a whole = 4.98 with 1 and 39 degrees of freedom.  The probability of a greater F given the null hypothesis that all effects are zero is 0.0315.

---

Dependent Variable DIFFPROC

---

| Variable | Coefficient | t-statistic with 53 degrees of freedom | Probability > \|t\| |
|---|---|---|---|
| CoopProc | 3.94 | 2.24 | 0.029 |
| Intercept | − 2.71 | − 3.81 | 0.000 |

---

$R^2 = 0.0866$;  Adjusted $R^2 = 0.0693$
F for the equation as a whole = 5.02 with 1 and 53 degrees of freedom.  The probability of a greater F given the null hypothesis that all effects are zero is 0.0292.

---

Juxtaposing the evidence in Tables 7.1 through 7.4 with the finding that a company's background and process environmental R&D is significantly greater when the company has cooperative R&D, there is evidence to support the claim that cooperative environmental R&D promotes efficiency.  Cooperative activity is especially likely for companies with low to moderate research intensity, for companies facing difficult appropriability conditions, and for companies facing great risks. For process R&D, ventures are less likely when spillovers (SPILLC) are important, and that does suggest that cooperation may reflect attempts to share knowledge that would not otherwise be available.  The effect is not significant by conventional standards, however.  Further, there is no evidence that seller concentration has a strong effect on the probability of cooperation one way or the other.  If anything, companies operating in concentrated industries are less likely to join cooperative ventures as illustrated in one of the specifications for basic research.  But the effect is just a very marginal effect at best and is never significant by conventional standards. Companies in the highly concentrated industries tend to spend more on environmental R&D as we saw in Chapter 6. And the companies in cooperative ventures spend more on environmental R&D, other things being the same. In all, the facts support the conventional views of efficiency from cooperation and do not suggest that cooperation is being used as a way to lessen environmental R&D by avoiding competition.

Firms may be groping in a Nelson and Winter (1982) world, but taken together the facts support the view that the process is efficiency enhancing nonetheless. Companies appear to be cooperating to realize efficiencies in some broad sense, and there is no indication that the cooperation eliminates desirable competition. This view of cooperative ventures is broadly consistent, then, with the Link and Bauer (1989) view of research joint ventures.

My assessment of these cooperative investments designed to improve emissions performance is quite different from my assessment (Scott, 1988) of the early cooperative ventures filed under the NCRA. For those ventures, I concluded (p. 183):

> What we are seeing with the NCRA co-operative R&D appears to be very similar to diversification behavior except that some competition is eliminated. The co-operative R&D protected by the NCRA has occurred in industries that were, during the 1970s, concentrated, with higher productivity growth and having R&D activities purposively combined by diversified firms with R&D in other industries. Also, co-operative R&D has not been more prevalent in those industries for which Levin *et al.* (1984) found appropriability difficulties, and therefore the act does not appear to be fostering R&D where competing firms dared not invest because of appropriability problems. Further, co-operative R&D appears to be more likely in industries where diversified firms were already investing relatively heavily, and to be less likely in those industries where they had low R&D intensity. Finally, broad areas of R&D investment combined by the co-operative R&D projects protected by the NCRA in the mid-1980s parallel closely the areas combined by the diversified firms of the mid-1970s. It is then plausible that the NCRA will stimulate cooperative projects that . . . reduce the net social benefit of R&D investment.

There are many potential reasons for the somewhat different assessments and the somewhat different questions addressed. Among them: (1) the cooperative investments studied here are for 1992–3, while those studied earlier were for 1985–6; (2) the data here are for emissions-reducing R&D, while the early information was for R&D on projects of all types; (3) the level of observation in the present study is the individual company, and I describe the R&D project, the company, and the industry characteristics that influence the company's decision to enter a cooperative research venture, while the level of observation in the earlier study was the industry and I compared the characteristics of industries with cooperative R&D with those without such R&D; (4) in part because of the differing levels of observation, the earlier study and the present one pose quite differently the descriptive question of whether the ventures essentially eliminate desirable competition and reduce R&D or instead increase R&D by improving appropriability conditions for investments where R&D and productivity had been lagging; (5) perhaps firms with 'good' reasons for entering joint research ventures or proud of their participation were more likely to reply to my survey although controlling for selection into my sample does not affect my conclusions; (6) it could be that companies have learned how to cooperate over time; and finally, (7) my earlier study sampled only NCRA ventures while the present study admitted all cooperative emissions-reducing ventures that the responding companies reported.

In all, the evidence in this chapter along with the evidence in Chapter 6 supports the view that the government has the power to increase desirable and important emissions-reducing research. First, the review of the NCRA ventures before and after the CAAA supports the view that companies will turn their cooperative research toward emissions problems that have become the focus of government concern. Second, the results about emissions-reducing R&D support the view that the companies will invest in research in order to reduce the probability of failing to meet the emissions standards imposed by government. My earlier (Scott, 1988) study at the industry level of observation of NCRA cooperative ventures concluded that there was evidence that such ventures might reduce desirable competitive pressures and hence eliminate socially desirable R&D investments. The present study was able to use primary data at a highly disaggregated level of observation — namely, observations for individual companies of research efforts of a particular kind. For that particular type of research, these new 'below the company level' data do not suggest that cooperation reduces desirable competitive pressures. Instead, cooperation appears to foster new research that would not have been initiated without the cooperation among companies.[4]

---

[4] FM Scherer observed (personal correspondence) that my finding for the environmental R&D of the 1990s contrasts with earlier US evidence (in the US Congressional Record for May 18, 1971, p. H4063) that the automobile industry used cooperative R&D to delay environmental improvements.

# 8. Environmental R&D, Emissions, and Policy

Chapter 1 presented the data from the Environmental Protection Agency (EPA) that shows the large reduction in toxic releases from 1988 through 1999 for the original set of reporting industries — the twenty two-digit manufacturing industries — and for the set of chemicals with reports throughout the period. The large reduction in toxic releases is reasonably hypothesized to be a result in part of the investments that industry has made in environmental R&D. In this concluding chapter, we shall test the hypothesis that environmental R&D has led to reductions in toxic releases in industry. We shall also ask whether environmental R&D and the pace of emissions reductions have changed over the last decade. Finally, the chapter concludes with a recommendation for an aggressive public policy to encourage environmental R&D.

## ENVIRONMENTAL R&D AND EMISSIONS REDUCTIONS

For the core set of chemicals traced since 1988, Table 8.1 presents EPA data (US EPA, 2001, Table 5-7, p. 5-17). The table shows, for each two-digit SIC industry, the variable RDCTN8899, the percentage reduction in total on-site and off-site releases over the period from 1988 to 1999 — the full period for which EPA has traced the releases of toxic chemicals for those original reporting industries. Table 8.1 also shows our estimates of ENVRD/SALES for each two-digit industry for which the necessary data were available. For a two-digit SIC industry, ENVRD/SALES is the estimated ratio of environmental R&D to sales for the companies in the industry.

The estimate of ENVRD/SALES for each two-digit SIC manufacturing industry was derived in four steps. First, the model in Table 4.2 (with the hazard rate added to that specification) of Chapter 4 was used to predict the ratio of environmental R&D to total R&D for each of the 71 companies that provided detailed reports about their environmental R&D in their response to the survey.

Second, for each of the 71 companies, that ratio was multiplied by the company's R&D to have an estimate of its environmental R&D. Third, each company's estimated environmental R&D was then divided by its sales to get its environmental R&D intensity. Fourth, the 71 companies were grouped by their primary two-digit manufacturing industry (15 industries were represented), and the environmental R&D intensities for each group were averaged to have the 15 two-digit industry environmental R&D intensities shown in Table 8.1. These are surely very rough estimates of the actual environmental R&D intensities for the industries, but they provide an interesting exploratory look at the relation between environmental R&D and emissions reductions across industries.

*Table 8.1. Environmental R&D Intensity and Reduction in Toxic Releases*

| SIC Industry | ENVRD/SALES | RDCTN8899 |
|---|---|---|
| 20 | 0.00332 | – 62.0 |
| 21 | . | 14.2 |
| 22 | 0.00455 | 78.5 |
| 23 | . | 73.3 |
| 24 | 0.00126 | 0.4 |
| 25 | 0.000967 | 75.6 |
| 26 | 0.0025 | 16.5 |
| 27 | . | 63.8 |
| 28 | 0.00882 | 56.3 |
| 29 | 0.00152 | 50.4 |
| 30 | 0.00189 | 41.9 |
| 31 | . | 62.4 |
| 32 | . | 27.8 |
| 33 | 0.00187 | 8.2 |
| 34 | 0.00589 | 53.2 |
| 35 | 0.00695 | 81.7 |
| 36 | 0.0215 | 76.7 |
| 37 | 0.00623 | 54.0 |
| 38 | 0.00997 | 89.7 |
| 39 | 0.00714 | 69.8 |

These statistical measures are very rough, and surely they are biased against finding the hypothesized relation between environmental R&D and the reduction in toxic releases. Yet the hypothesized relationship is clearly present in the data. Table 8.2 estimates the simple model linking the two variables by the elasticity of the percentage reduction in an industry's toxic releases with respect to the environmental R&D intensity for the industry.

Over the period from 1988 through 1999, all two-digit manufacturing industries except for SIC 20, the food industry, experienced a drop in toxic releases. The food industry showed a large increase until a large drop during 1998 through 1999. It is clearly an outlier, and we shall examine the data with that in mind. Table 8.2 presents two specifications. The first specification drops the observation for the food industry and estimates the simple model $RDCTN8899 = \alpha(ENVRD/SALES)^\beta e^\varepsilon$, where $e$ is the base for the natural logarithms and $\varepsilon$ is random normal error. The second specification keeps the observation for the food industry, but assigns it a percentage reduction of 1, with the natural logarithm of 0, and then includes a dummy variable for the food industry. The dummy dFOOD equals 1 for the observation for industry SIC 20, and it equals zero otherwise.

*Table 8.2.  OLS Model for ln RDCTN8899 and ln (ENVRD/SALES)*

Dependent Variable ln RDCTN8899 and without the Food Industry

| Variable | Coefficient | t-statistic with 12 degrees of freedom | Probability > \|t\| |
|---|---|---|---|
| ln(ENVRD/SALES) | 0.816 | 2.09 | 0.059 |
| Intercept | 8.065 | 3.68 | 0.003 |

$R^2 = 0.266$;  Adjusted $R^2 = 0.205$
F for the equation as a whole= 4.35 with 1 and 12 degrees of freedom.  The probability of a greater F given the null hypothesis that all effects are zero is 0.0591.

Dependent Variable ln RDCTN8899 except for Food Industry which is assigned the value 0 for ln RDCTN8899

| Variable | Coefficient | t-statistic with 12 degrees of freedom | Probability > \|t\| |
|---|---|---|---|
| ln(ENVRD/SALES) | 0.816 | 2.09 | 0.059 |
| dFOOD | − 3.41 | − 2.54 | 0.026 |
| Intercept | 8.065 | 3.68 | 0.003 |

$R^2 = 0.487$;  Adjusted $R^2 = 0.402$
F for the equation as a whole= 5.70 with 2 and 12 degrees of freedom.  The probability of a greater F given the null hypothesis that all effects are zero is 0.0182.

Table 8.3 uses ordinary least squares to reestimate the relation between the reduction in emissions and environmental R&D. The first specification in the table drops the food industry, and the second specification keeps it.

*Table 8.3. OLS Model for RDCTN8899 and (ENVRD/SALES)*

Dependent Variable RDCTN8899 and without the Food Industry

| Variable | Coefficient | t-statistic with 12 degrees of freedom | Probability > \|t\| |
|---|---|---|---|
| ENVRD/SALES | 2723.8 | 2.12 | 0.056 |
| Intercept | 38.01 | 3.79 | 0.003 |

$R^2 = 0.272$; Adjusted $R^2 = 0.211$
F for the equation as a whole = 4.48 with 1 and 12 degrees of freedom. The probability of a greater F given the null hypothesis that all effects are zero is 0.0558.

Dependent Variable RDCTN8899 and with the Food Industry

| Variable | Coefficient | t-statistic with 13 degrees of freedom | Probability > \|t\| |
|---|---|---|---|
| ENVRD/SALES | 3372.6 | 1.76 | 0.102 |
| Intercept | 27.09 | 1.86 | 0.085 |

$R^2 = 0.192$; Adjusted $R^2 = 0.130$
F for the equation as a whole = 3.09 with 1 and 13 degrees of freedom. The probability of a greater F given the null hypothesis that all effects are zero is 0.102.

The essential results are the same for all the specifications. The data, although crude and approximate, support the hypothesis that environmental R&D is associated with the reduction in toxic releases. Figure 8.1 illustrates the relationship from Table 8.2 fitted through the scatter diagram for the two variables.

Next we shall address the evidence about trends in the two variables — environmental R&D and the reduction in the releases of toxic chemicals. The evidence available suggests that on the whole the two variables have fallen somewhat. Relying on the confidence intervals, however, the conservative conclusion is simply that neither environmental R&D nor the reduction is toxic releases have increased in any pronounced way. First, the EPA data show that overall the average annual reductions in releases have fallen somewhat. Second, a survey of manufacturing firms shows that on average the proportion of total company R&D going to environmental R&D has also fallen. For both variables, however, there is considerable variance around the average.

*Figure 8.1.  Illustration of the Fitted (∆) and Actual (o) Values for ln RDCTN8899 as a Function of ln (ENVRD/SALES)*

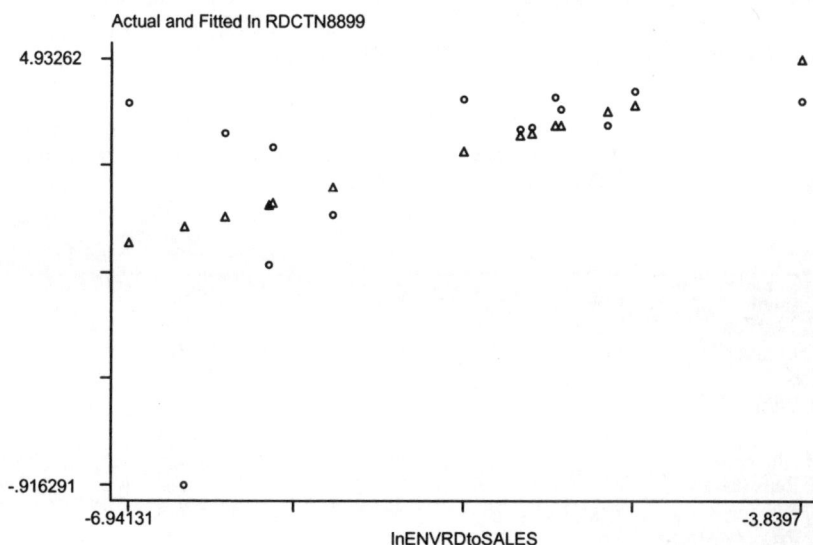

Actual and Fitted ln RDCTN8899

4.93262

-.916291

-6.94131                                                    -3.8397

lnENVRDtoSALES

Turning first to the EPA data, on average for the 11 years from 1988 through 1999, the annual percentage reduction in toxic releases for each two-digit industry is given by AVGANN8899 = RDCTN8899/11.  The average across the 20 two-digit industries of that annual percentage reduction is 4.24.  EPA (US EPA, 2001, Table 5-6, p. 5-15) also provides the percentage reduction in toxic releases for each two-digit manufacturing industry over the period from 1995 through 1999.  Dividing that reduction by 4 gives that annual percentage reduction AVGANN9599 for each industry during those years.  Averaging across the 20 two-digit industries shows an average of 3.83 for the annual percentage reduction for the recent period.  There is considerable variance across the two-digit industries in the annual percentage reduction for the entire period and for the most recent period, and the variance is especially large for the recent period.

For the averages across the 20 industries, the 95 percent confidence interval (0.149 to 7.51) for the average annual percentage reduction in the recent period includes the confidence interval (2.68 to 5.80) for the measure for the entire period.  In that sense, although the arithmetic average is lower in the recent period, the results for the entire period are not very different from the results for the recent period.  Given all of the caveats that EPA makes about the data — caveats that were briefly reviewed in Chapter 1 — about all one can say is that the annual reductions in toxic releases in manufacturing industry do not, on the whole, appear to have increased.

*Table 8.4. Industry Coverage for the Follow-up Sample*

| Industry | Respondents in Original Sample with the Industry as Primary Industry | Respondents in Follow-up Sample with the Industry as Primary Industry | Respondents in Follow-up Sample with the Industry as a Secondary Industry |
|---|---|---|---|
| D20 Food | Yes | Yes | No |
| D22 Textiles | Yes | Yes | Yes |
| D24 Lumber & Wood | Yes | No | Yes |
| D25 Furniture | Yes | No | Yes |
| D26 Paper | Yes | No | Yes |
| D28 Chemicals | Yes | Yes | Yes |
| D29 Petroleum | Yes | No | Yes |
| D30 Rubber | Yes | No | Yes |
| D33 Primary Metals | Yes | No | Yes |
| D34 Fabricated Metals | Yes | Yes | Yes |
| D35 Industrial Machinery | Yes | Yes | Yes |
| D36 Electronics | Yes | Yes | Yes |
| D37 Transportation | Yes | Yes | Yes |
| D38 Instruments | Yes | Yes | Yes |
| D39 Miscellaneous Manufacturing | Yes | No | No |
| Nonmanufacturing | Yes | Yes | Yes |

To gain some insights into how environmental R&D may have changed over the last decade, I contacted a representative sample of the 71 companies that provided Chapter 4's detailed information about environmental R&D in the early 1990s. I was able to gain the insights of the R&D directors or environmental affairs directors of 17 of those 71 companies. The 17 firms provided good coverage of the industries represented in the original sample described in Chapter 3.

Table 8.4 compares the original coverage with the coverage of industries by the 17 firms that in 2001 provided an update on their environmental R&D. As explained in Chapter 3, the original survey had no respondents with primary industry SIC 21 (tobacco), SIC 23 (apparel), SIC 27 (printing), SIC 31 (leather), or SIC 32 (Stone, Clay, and Glass). Otherwise, the two-digit manufacturing SIC industries and nonmanufacturing industry were represented by firms with one of the industries as their primary industry. Of the original industries covered, all but SIC 39 (miscellaneous manufacturing) were represented by firms in the follow-up sample.

The finding from the 17 firms is that on average the proportion of total R&D that is taken by environmental R&D has fallen somewhat. For the 17 companies, that proportion is on average 92.65 percent of what it was a decade ago. There is a large variance in the responses, with some respondents reporting that they have abandoned environmental R&D and others reporting that as a proportion of their total R&D they have doubled it. For these 17 companies, the ratio of environmental R&D to total industrial R&D is 92.65 percent of what it was a decade ago, but the 95 percent confidence interval for that figure ranges from 65.41 percent to 119.89 percent. Conservatively, the most we can say with the data in hand is that on the whole the relative amount of environmental R&D in industry has not increased significantly during the last decade. Because the environmental R&D during the 1990s has reduced industrial environmental problems, perhaps no increase in the relative importance of environmental R&D is acceptable. However, many serious environmental problems remain, and to some extent new research is needed not only to solve those outstanding problems, but to avoid creating new problems as the processes and products from industry evolve.

## PUBLIC POLICY

The preceding chapters have made the case that public concern about toxic emissions causes companies to invest in R&D to develop new processes and products with better environmental performance. Chapters 3 and 4 describe the extent and types of that R&D, and Chapters 1 and 2 describe the emissions problems addressed by the R&D. Chapter 5 describes the theory linking public concern to environmental R&D, and Chapters 6 and 7 test hypotheses from the theory. In this chapter, we have shown that environmental R&D is associated with the reduction in toxic releases across industry. Industry's toxic releases into the environment are nonetheless still substantial. Although the last decade has shown environmental R&D to be effective at reducing emissions, *on average* environmental R&D intensity and the pace of emissions reduction have fallen.

I conclude with a recommendation for an aggressive public policy to increase environmental R&D. Such R&D is substantial, yet the *Toxics Release Inventory* described in Chapter 1 shows there is still an immense amount of industrial toxic waste. Rather than simply continue to look for ways to dispose of it, focused R&D offers the hope of new processes and products that will have less toxic wastes as their byproducts. Further, our simulations of the model show that environmental R&D varies greatly with variables reflecting the pressures to do R&D *other than* the extent of the toxic emissions problems. For example, in certain circumstances more R&D is done when companies face greater Schumpeterian competitive pressures, holding constant the effect of the emissions problems. That implies that the socially optimal amount of environmental R&D does not occur in at least some — and one suspects in many and perhaps all — industries.

If government decides that further reductions in toxic byproducts of industry are desirable, then policy instruments that increase environmental R&D are useful. The theoretical models of Chapter 5 imply two straightforward policies to increase environmental R&D, as long as the policies are not so severe that there is no profit left in operating at all. One policy is a periodic preinnovation tax. The tax would be paid until a firm developed or acquired innovations that achieved desired environmental performance.[1] The other policy that follows from the theories of Chapter 5, and that increases environmental R&D, is a policy of tougher environmental standards. We can explain the effects of these two policies with both of Chapter 5's models of the environmental R&D decision.

First, consider again the uncomplicated model. From Chapter 5, the equilibrium investment in R&D is:

$$\ln RIII = \frac{-\ln a - \ln(1-\lambda) - \ln(-\beta_R) - \ln A}{\beta_R - 1}$$

$$+ \left(\frac{-b}{\beta_R - 1}\right) \ln SALES + \sum_{i}^{n} \left(\frac{-\beta_i}{\beta_R - 1}\right) \ln X_i \, .$$

In this simplest model, a periodic preinnovation tax affects $\lambda$, the proportion of company value when environmental regulations are met that remains when performance falls short of mandated performance. The periodic preinnovation tax would mandate an innovation that eliminated toxic waste to some specified extent. Hence, the preinnovation tax would lower $\lambda$, because the tax is paid in every period until the mandated innovation occurs. The partial derivative of environmental R&D with respect to $\lambda$ is negative, since $\partial \ln RIII / \partial \lambda = (1/(\beta_R - 1))(1/(1-\lambda)) < 0$. Thus, imposing or increasing the tax lowers $\lambda$ and that causes the company to increase RIII.

---

[1] The periodic preinnovation tax is developed in the context of a general R&D model in Scott (1995). In this chapter, we explain the tax in the context of Chapter 5's two models of the environmental R&D decision. Scott (1995) also discusses and compares current public policy toward R&D, including the more conventional taxation policy of tax credits for R&D investments.

Further, in the simplest model, if $X_p$ is a policy instrument that has a positive sign in its effect — $\beta_p > 0$ — on the probability that process technology or product performance will not meet regulatory muster, then increasing the setting of the instrument will cause the company to increase its environmental R&D. The stringency of the standard is just such a policy instrument. Hence, making the standard more stringent will increase the amount of environmental R&D that the company finds profitable.

Now (recalling that the notation is redefined for this second model — see Chapter 5), consider the more detailed model for which the equilibrium R&D investment is:

$$\ln RIII = \left( \frac{1}{(1-(\beta_R \ln(\frac{1}{2})))} \right) \left( \begin{array}{l} \ln a + 2\ln(\frac{1}{2}) + \ln(\frac{1}{2})\ln A + \ln \beta_R + b\ln SALES \\ +\sum_i^m b_i \ln Z_i + \sum_i^n (\ln(\frac{1}{2}))\beta_i \ln X_i \end{array} \right).$$

In Chapter 5, we observed that the effect of a value-shifting variable $Z_i$ has the sign for the elasticity ($b_i$) for the model's R&D value parameter with respect to the variable.[2] The periodic preinnovation tax is one of these value-shifting $Z_i$ variables. The elasticity of R&D value with respect to the tax is positive, because tax must be paid until the mandated innovation occurs. R&D increases the probability of an innovation with adequate environmental performance; hence, it increases the probability of avoiding the tax — realizing the value of not having to pay it. The tax increases the value of doing R&D.

In the model, the X variables determine alpha; as alpha increases there is a favorable shift in the probability distribution over the performance — as judged relative to the regulated environmental standards — for the innovative technology produced by a company's environmental R&D investments. As alpha increases, the distribution shifts to the right, centering over higher performance for the innovative technology. In Chapter 5, we found that for a variable that makes alpha smaller as the variable is increased, then increasing the variable will cause a company to do more environmental R&D. The company, other things being the same, will find it profitable to do more R&D to make the distribution for the performance of its innovation more favorable. The stringency of the standard for acceptable environmental performance is just such a variable. The more stringent the standard as measured by an index variable X, then the smaller alpha is. Industry is given a tougher environmental problem to solve, and R&D will be greater.

With the two policies, government has the opportunity for an aggressive policy of imposing preinnovation taxes in conjunction with tougher standards. Environmental R&D will increase in response to the policies, as long as the tax

---

[2] The R&D value parameter $\gamma$ is modeled in Chapter 5. The parameter reflects the value of good environmental performance, and that good performance is obtained with successful environmental R&D.

and the standard are not so severe that profits are negative given the optimal levels of production and R&D.

We have developed in the earlier chapters a description of the portion of industrial R&D aimed at solving environmental problems. However, much work in the area of environmental problems is done with routine engineering that moves companies along a learning curve for essentially the same technologies or perhaps, in some cases, actually develops new technologies. The key, though, is that — as Hollander (1965) found for Du Pont's rayon-process innovation — many companies do this within their engineering budgets, rather than with resources included in their R&D budgets. Several of the respondents spoke to this issue, some with written comments, and some with telephone conversations with me.

For example, one executive of a large consumer products company wrote: 'We try to avoid recognized toxins, to the degree possible, in our formulation and processes. Your use of the term research caused me conceptual problems. We do substantial research on the safety of our products and raw materials. . . . . We are a manufacturing company, and as such, we attempt to put our engineering and process development efforts to minimizing the use or creation of all toxic release from our process design. I characterize this as engineering, not research in the context we think of it at [company name].'

Evidently, 'research,' in the sense of the formal R&D budget, at this company comes up with new products, and then the engineers, sensibly, are expected to design processes for producing those products with a minimum of toxic emissions. Of course, some of the related engineering expenses may well be R&D investment in the sense that I would like to measure it. The 'gray area' where routine engineering and R&D investment overlap may be particularly important for the R&D for air toxic emissions that I have studied in this book.

For another example, a manufacturer of industrial machinery and equipment reports: 'We are working on product redesign as well as working with degreasing agent suppliers to eliminate 1,1,1 TCE [1,1,1-Trichloroethane] from our manufacturing processes.' The manufacturer, however, does not include this work in its R&D budget, but instead includes it under routine engineering.

Despite the fact that apparently much is done to clean up air toxic emissions with routine engineering, R&D provides a different approach that is not simply moving down a learning curve for known processes or relegating new developments to routine engineering. Doing R&D, the company is focused on risky investments to develop new approaches, and it puts into play a part of the corporation with its identity and success bound up with the development of those new approaches rather than with incremental improvements to existing processes. The concerns about Title III chemicals and environmental problems more generally warrant the focus of our corporations' R&D apparatus.

The Title III chemicals have generated considerable concern about public health. In Chapter 2, we have seen that concern reflected in the medical journals. The Clean Air Act Amendment itself in describing Title III chemicals (US Congress, 1990, pp. 2535–6) cites the 'threat of adverse human health effects' from 'substances which are known to be, or may reasonably be anticipated to be, carcinogenic, mutagenic, teratogenic, neurotoxic, which cause reproductive dysfunction, or which are acutely or chronically toxic.' R&D to reduce those emissions could have a big payoff, and the results of this study suggest ways to increase the R&D aimed at reducing toxic air emissions and environmental problems more generally.

The costs of the negative externalities caused by toxic air emissions are unknown but believed to be large. Public concern about the toxic emissions has led to public pressure such as legislated in the Clean Air Act Amendments (US Congress, 1990). Both the theory and the estimation in this book have shown that if companies are worried about meeting regulatory standards that restrict toxic emissions, then the companies will invest more in R&D to ensure that the standards are met. Successful R&D investments are typically fraught with uncertainty, and coping with such uncertainty is something that most companies would reasonably prefer to avoid.[3] In that context, where society wants to reduce the negative externalities in production, the role for regulation is to set standards that cause firms to accept the uncertainty of R&D investment as a necessary part of corporate existence.[4] Further, we have seen that the desired R&D-increasing pressure could come not only from more stringent standards, but also from a tax that must be paid until successful innovation occurs.

Advocates for environmental protection have expressed concern that the administration of President George W. Bush will 'roll back or weaken various environmental regulations . . . . Environmentalists fear that the Clean Air Act might be hurt most by Bush's policies.' (Shogren, 2001, p. A1) The evidence gathered and interpreted in this book shows that public pressure to reduce emissions interacts with the profit-maximizing behavior of firms to increase R&D investment in emissions-reducing technology. Thus, the greatest cost of relaxing the constraints of the Clean Air Act may be the reduction in environmental R&D — R&D that has been stimulated by the law — and the consequent reduction in innovations that improve the environmental performance of industry.

---

[3] Klein (1977) organizes his views about technological change around the theme that the acceptance of an uncomfortable and unstable state of uncertainty is a concomitant of innovation.

[4] Our model in which environmental regulation increases R&D is consistent with the stronger interpretations of the 'Porter hypothesis' (Jaffe *et al.*, 1995, pp. 153–7) in which environmental regulation actually increases productivity, growth, and competitiveness.

# Glossary for Variables[1]

APP. AppBack, AppProc, and AppProd are defined respectively for companies reporting background emissions research, process R&D to reduce emissions, and product R&D to reduce emissions. For companies with the particular type of research, the APP variable equals 1 if the company reported that it was difficult or very difficult to realize a normal return from its investments in the research or R&D. Otherwise the variable equals 0. Thus, the variable is defined as 1 for those companies choosing the third or the fourth response to questions #6, #19, or #32, and as 0 for those companies choosing the first or second response to those questions.

BACK3. An estimate of the proportion of a company's R&D devoted to background research on Title III toxic air emissions.

BACK3RD. An estimate of a company's background research on Title III emissions, in millions of 1992 dollars.

BACKOTHER. A dummy variable that takes the value 1 if a company's background research on emissions received additional financing from other companies or from the government; otherwise = 0.

COOP. CoopBack equals 1 if a respondent chose the first answer to survey question #5 — i.e., if the company said that it had background emissions research that was performed in a cooperative venture with other firms. If a company had background emissions research but did not perform any of that research in cooperation with other firms, then CoopBack equals 0. CoopProc equals 1 if a respondent chose the first answer to survey question #18 — i.e., if the company said that it had process R&D to reduce emissions that was performed in a cooperative venture with other firms. If a company had process R&D to reduce

---

[1] At times the definitions of the variables refer to the questions in the survey. The appendix to Chapter 3 provides the complete survey with the numbered questions.

emissions but did not perform any of that R&D in cooperation with other firms, then CoopProc equals 0. CoopProd equals 1 if a respondent chose the first answer to survey question #31 — i.e., if the company said that it had product R&D to reduce emissions that was performed in a cooperative venture with other firms. If a company had product R&D to reduce emissions but did not perform any of that R&D in cooperation with other firms, then CoopProd equals 0.

COST. CostBack, CostProc, and CostProd are defined respectively for companies reporting background emissions research, process R&D to reduce emissions, and product R&D to reduce emissions. For companies with the particular type of research, the COST variable equals 1 if the company reported that as compared with its other research or R&D projects the emissions work was more costly. Otherwise the variable equals 0. Thus, the variable is defined as 1 for those companies choosing the third response to questions #8, #21, or #34, and as 0 for those companies choosing the first or second response to those questions.

CR4C. Measures average seller concentration for a company's industries. The concentration ratio of the value of industry shipments is CR4, the four-firm seller concentration ratio as a percentage for each four-digit manufacturing SIC industry (US Department of Commerce (1992, Table 4, pp. 6.4 – 6.45)). Then, for each company CR4C is the average value of CR4 across the four-digit manufacturing industries in which a company operates.

DBACKGROUND. A dummy variable = 1 if a respondent reported background research on emissions; otherwise = 0.

DIFF. Using our econometric models, we make predictions of the three types of Title III R&D — BACK3RD, PROC3RD, and PROD3RD — for each of the firms that *have* each particular type of Title III R&D. We can then measure the difference DIFF between the actual amount of each type of R&D from its predicted amount (the conditional expectation given RIII > 0) for those firms having each type of R&D. Then, for just those firms having the particular type of Title III R&D, we have the differences DIFFBACK, DIFFPROC, and DIFFPROD for background research on Title III emissions, Title III process R&D, and Title III product R&D respectively.

DPROCESS. A dummy variable = 1 if a respondent reported process R&D to reduce emissions; otherwise = 0.

DPRODUCT. A dummy variable = 1 if a respondent reported product R&D to reduce emissions; otherwise = 0.

D20, D21, . . ., D39. Dummy variables indicating the 20 two-digit SIC manufacturing industries. Each of these variables is defined in Chapter 3 in the table showing the primary industry effects on response, and then used in other empirical specifications throughout the book.

DR. A dummy that equals 1 if the firm was one of the 150 that responded by answering the original questionnaire in 1993 and is 0 otherwise.

EEI. The Investor Responsibility Research Center's 'emissions efficiency index' which shows, for the three years 1988–90, the ratio of reported toxic chemical emissions in pounds to the company's domestic revenues expressed in thousands of dollars.

EFF. EffBack, EffProc, and EffProd are defined respectively for companies reporting background emissions research, process R&D to reduce emissions, and product R&D to reduce emissions. For companies with the particular type of research, the EFF variable equals 1 if the company reported that, as compared with its other research or R&D projects, efficiencies from the size of the research effort were more important for the emissions work. Otherwise the variable equals 0. Thus, the variable is defined as 1 for those companies choosing the third response to questions #10, #23, or #36, and as 0 for those companies choosing the first or second response to those questions.

ENVRD/SALES. Our estimate for each two-digit SIC industry of the ratio of environmental R&D to sales for the companies in the industry.

Envtotot. Of the 71 respondents reporting R&D to reduce air-emissions problems, 47 of the companies had consistent answers across the ways of calculating the ratio of environmental to total R&D for the company. The variable Envtotot is for each of those 47 companies the consistent answer given for that ratio.

Hazard Rate. The hazard rate from Chapter 3's probit model for response is used to control for selection into the sample as discussed in Chapters 4, 6, and 7.

HIGH. A dummy variable that equals 1 for the two-digit industry categories with high R&D intensity for the firms in our sample. Chapter 7 provides details.

IMPSC. Measures average import competition faced by the company in its industries. The measure of import competition is IMPS, the ratio of imports to shipments for each four-digit manufacturing industry. Then, for each company IMPSC is the average value of IMPS across the four-digit manufacturing industries in which a company operates. The measure IMPS is constructed from information in US Department of Labor (1991) and in US Department of Commerce (1990). To get imports for each four-digit manufacturing SIC industry using the new 1987 SIC industries, I began with US Department of Labor (1991, Table 1, pp. 3–8). This was the first issue of *Trade & Employment* that used the new 1987 industry categories. Although the issue focused on 1990, import data on the new SIC basis was also given for 1988. The import data are often for combinations of the 1987 four-digit industries, or for pieces of those industries. I used US Department of Commerce (1990, Table 1, pp. 2.5 – 2.32) to get the value of product shipments for SIC four-digit manufacturing industries and for the five-digit parts of those four-digit industries. These shipments were then used to allocate the imports among the SIC four-digit industries. For example, if an import total was given for two or more four-digit industries, the imports were allocated to the industries in the proportion of those industries' shipments. In many cases, the imports figure was for a combination of five-digit industries. For those cases, the five-digit shipments were used to allocate the imports among the five-digit industries. After all such imputations were made, the imports for the five-digit industries were then combined into their appropriate four-digit industries to get the final figure for imports for the four-digit industry.

LOW. A dummy variable that equals 1 for the two-digit industry categories with low R&D intensity for the firms in our sample. Chapter 7 provides details.

MED1, MED2, MED3. Alternative measures of the number of medical journal articles studying a chemical. See Chapter 2 for details.

MED2PR. The average of MED2 for all of the Title III chemicals associated with a company's primary four-digit industry.

MODERATE. A dummy variable that equals 1 for the two-digit industry categories with moderate R&D intensity for the firms in our sample. Chapter 7 provides details.

N4DIG. The number of four-digit SIC manufacturing industries that handle, treat, or produce a chemical during the manufacturing process.

NTAP. The number of Title III toxic air pollutants associated with a four-digit SIC manufacturing industry.

NTAPC. Measures the extent of a company's Title III emissions problems. NTAPC is the average value of NTAP, the number of Title III toxic air pollutants associated with a manufacturing industry, across the four-digit manufacturing SIC industries in which the company operates.

NTAP2. The total number of toxic air pollutants associated with a two-digit SIC manufacturing industry.

NTAP2.III. The number of Title III toxic air pollutants associated with a two-digit SIC manufacturing industry.

N[TAP,4D]2. The total number of 'emissions' associated with a two-digit industry, where an 'emission' is defined as an instance of a toxic air pollutant being associated with a four-digit industry.

N[TAP,4D]2.III. The number of Title III 'emissions' associated with a two-digit industry, where an 'emission' is defined as an instance of a toxic air pollutant being associated with a four-digit industry.

PROC3. An estimate of the proportion of a company's R&D devoted to process R&D to reduce Title III air emissions.

PROC3RD. An estimate of a company's Title III process R&D, in millions of 1992 dollars.

PROCOTHER. A dummy variable that takes the value 1 if a company's process R&D to reduce emissions received additional financing from other companies or from the government; otherwise = 0.

PROD3. An estimate of the proportion of a company's R&D devoted to product R&D to reduce Title III air emissions.

PROD3RD. An estimate of a company's Title III product R&D, in millions of 1992 dollars.

PRODOTHER. A dummy variable that takes the value 1 if a company's product R&D to reduce emissions received additional financing from other companies or from the government; otherwise = 0.

RD. Is the company's R&D expenses, which are stated in millions of dollars, for the most recent fiscal year as of May 18, 1993, as reported in *Business Week* (1993).

RDCTN8899. For the core set of chemicals traced since 1988, EPA data (US EPA, 2001, Table 5-7, p. 5-17) provides, for each two-digit manufacturing SIC industry, the variable RDCTN8899. The variable is the percentage reduction in total on-site and off-site releases over the period from 1988 to 1999 — the full period for which EPA has traced the releases of specified toxic chemicals for those original reporting industries.

RISK. RiskBack, RiskProc, and RiskProd are defined respectively for companies reporting background emissions research, process R&D to reduce emissions, and product R&D to reduce emissions. For companies with the particular type of research, the RISK variable equals 1 if the company reported that the work was more risky in comparison with the riskiness of its other research or R&D projects. Otherwise the variable equals 0. Thus, the variable is defined as 1 for those companies choosing the third response to questions #9, #22, or #35, and as 0 for those companies choosing the first or second response to those questions.

SALES. The measure of a company's size. For each company, SALES is measured as sales, for the company's most recent fiscal year as of May 18, 1993, in millions of dollars, as reported in *Business Week* (1993), and is contemporaneous with the R&D expenses also reported there.

SPILLC measures the potential for spillovers of R&D knowledge useful to the company as it pursues emissions-reducing R&D. One might expect that the spillovers for a company would be larger when the number of industries associated with the Title III chemicals of a company's typical line of business is larger. For each Title III chemical, we know the number N4DIG of four-digit SIC manufacturing industries that are associated with the chemical. For each four-digit industry, then, we can obtain the average $\overline{N4DIG}$ of N4DIG across the four-digit industry's Title III chemicals. Then we can compute for each company SPILLC which is the average of $\overline{N4DIG}$ across the four-digit industries in which the company operates. Thus, SPILLC characterizes an average line of business for a company; the larger SPILLC is, the larger the potential for spillovers in the company's Title III R&D.

# References

Baldwin, WL, and Scott, JT, *Market Structure and Technological Change* (London; New York: Harwood Academic, 1987).

Baumol, WJ, and Oates, WE, *The Theory of Environmental Policy*, Second Edition (Cambridge, England; New York: Cambridge University Press, 1988).

*Business Week*, 'R&D Scoreboard,' June 28, 1993, pp. 105–125.

Cohen, WM, and Levin, RC, 'Empirical Studies of Innovation and Market Structure,' *Handbook of Industrial Organization*, vol. II, edited by R Schmalensee and RD Willig (Amsterdam: North-Holland, 1989).

Dales, JH, *Pollution, Property, and Prices* (Toronto: University of Toronto Press, 1968).

Dun & Bradstreet, *Million Dollar Directory: America's Leading Public & Private Companies*, Series 1993, 1994.

Friday, L, and Laskey, R, *The Fragile Environment* (Cambridge, England; New York: Cambridge University Press, 1989).

Greene, WH, *Econometric Analysis*, Third Edition (Upper Saddle River, New Jersey: Prentice Hall, 1997).

Hollander, S, *The Sources of Increased Efficiency: A Study of Du Pont Rayon Plants* (Cambridge, Massachusetts: The MIT Press, 1965).

Investor Responsibility Research Center, Environmental Information Service, *1993 Corporate Environmental Profiles Directory* (Washington, DC, Investor Responsibility Research Center, 1993).

Jaffe, AB, Peterson, SR, Portney, PR, and Stavins, RN, 'Environmental Regulation and the Competitiveness of US Manufacturing: What Does the Evidence Tell Us?' *Journal of Economic Literature*, vol. 33, no. 1 (March 1995), pp. 132–163.

Klein, B, *Dynamic Economics* (Cambridge, Massachusetts: Harvard University Press, 1977).

Kohn, M, and Scott, JT, 'Scale Economies in Research and Development: the Schumpeterian Hypothesis,' *The Journal of Industrial Economics*, vol. 30, no. 3 (March 1982), pp. 239–249.

Levin, RC, Klevorick, AK, Nelson, RR, and Winter, SG, 'Survey Research on R&D Appropriability and Technological Opportunity: Part 1,' Working Paper, Yale University (1984).

Link, AN, 'Research Joint Ventures: Patterns from *Federal Register* Filings,' *Review of Industrial Organization*, vol. 11, no. 5 (October 1996), pp. 617–28.

Link, AN, and Bauer, LL, *Cooperative Research in US Manufacturing* (Lexington, Mass.: Lexington Books, D. C. Heath and Company, 1989).

Leyden, DP, and Link, AN, *Government's Role in Innovation* (Dordrecht; Boston; London: Kluwer Academic Publishers, 1992).

Lynch, SK, *Crosswalk/Air Toxic Emission Factor Data Base Management System User's Manual*, Version 1.2, EPA-450/2-91-028, October 1991.

Maddala, GS, *Limited Dependent and Qualitative Variables in Econometrics* (Cambridge, England: Cambridge University Press, 1983).

Martin, S, 'Spillovers, Appropriability, and R&D,' *Journal of Economics*, vol. 75, no. 1 (2002), pp. 1–32.

Montgomery, WE, 'Markets in Licenses and Efficient Pollution Control Programs,' *Journal of Economic Theory*, vol. 5 (December 1972), pp. 395–418.

Mood, AM, and Graybill, FA, *Introduction to the Theory of Statistics*, Second Edition (New York: McGraw-Hill, 1963).

National Library of Medicine, MEDLINE, July 1993.

Nelson, RR, and Winter, SG, *An Evolutionary Theory of Economic Change,* (Cambridge, Mass.: Harvard University Press, 1982).

Nisbet, EG, *Leaving Eden: To Protect and Manage the Earth* (Cambridge, England; New York: Cambridge University Press, 1991).

OneSource Information Services Inc., 'OneSource, CD/Corporate: US Private+,' compact disk with US public and private companies, quarterly updates throughout 1993 and 1994, Feldberg Library, Amos Tuck School of Business Administration.

Scherer, FM, *Industrial Market Structure and Economic Performance* (Chicago: Rand McNally & Company, 1970).

Scherer, FM, *Innovation and Growth: Schumpeterian Perspectives* (Cambridge, Massachusetts: The MIT Press, 1984).

Schmid, RE, 'Fewer Toxics Are Being Released, US Reports — More Generated by Industry,' Associated Press, *Valley News*, April 20, 1994, p. B1.

Schumpeter, JA, *Capitalism, Socialism, and Democracy* (New York: Harper, 1942).

Schumpeter, JA, *History of Economic Analysis* (London: George Allen & Unwin, 1954).

Scott, JT, 'Diversification versus Co-operation in R&D Investment,' *Managerial and Decision Economics*, vol. 9 (1988), 173–86.

Scott, JT, *Purposive Diversification and Economic Performance* (Cambridge, England; New York: Cambridge University Press, 1993).

Scott, JT, 'The Damoclean Tax and Innovation,' *Journal of Evolutionary Economics*, vol. 5, no. 1 (February 1995), pp. 71–89.

Scott, JT, 'Environmental Research Joint Ventures among Manufacturers,' *Review of Industrial Organization*, vol. 11, no. 5 (October 1996), pp. 655–79.

Scott, JT, 'Schumpeterian Competition and Environmental R&D,' *Managerial and Decision Economics*, vol. 18 (1997), pp. 455–69.

Shogren, E, 'War Shields Environmental Rollbacks,' *Los Angeles Times* as reprinted in *Valley News*, December 27, 2001, pp. A1, A8.

Standard & Poor's *Register of Corporations, Directors and Executives 1993* (New York: Standard & Poor's Corporation, 1993).

Standard & Poor's *Register of Corporations, Directors and Executives 1994* (New York: Standard & Poor's Corporation, 1994).

StataCorp, *Stata Statistical Software: Release 7.0* (College Station, Texas: Stata Corporation, 2001).

Tietenberg, T, *Emissions Trading: An Exercise in Reforming Pollution Policy* (Washington, DC: Resources for the Future, 1985).

US 98th Congress (1984) *National Cooperative Research Act of 1984*. Public Law 98-462.

US 103rd Congress (1993) *National Cooperative Production Amendments of 1993* (HR 1313, 'National Cooperative Research and Production Act of 1993'), *Antitrust & Trade Regulation Report*, 64, 688:1–2, 717–32.

US Congress, Public Law 101-549, 101st Congress, 'An Act to amend the Clean Air Act to provide for attainment and maintenance of health protective national ambient air quality standards, and for other purposes,' November 15, 1990, 104 Stat. 2399–2712.

US Department of Commerce, *Value of Product Shipments*, 1988 Annual Survey of Manufactures, M88(AS)-2, Bureau of the Census, Issued October 1990 (Washington, DC: US Government Printing Office, 1990).

US Department of Commerce, *Concentration Ratios In Manufacturing*, 1987 Census of Manufactures, MC87-S-6, Subject Series, Economics and Statistics Administration, Bureau of the Census, Issued February 1992 (Washington, DC: US Government Printing Office, 1992).

US Department of Commerce (1993) *Research & Development Consortia Registered under the National Cooperative Research Act of 1984.* Clearinghouse for State and Local Initiatives on Productivity, Technology, and Innovation, Office of Technology Commercialization.

US Department of Labor, *Trade & Employment*, Bureau of Labor Statistics, Issued September 1991 (Washington, DC: US Government Printing Office, 1991).

US Environmental Protection Agency (EPA), Office of Air Quality Planning and Standards, Research Triangle Park, North Carolina, *Crosswalk/Air Toxic Emission Factor Data Base Management System (XATEF)*, Version 1.2 (for microcomputers), EPA-450/4-91-028, October 1991.

US Environmental Protection Agency (EPA), *1999 Toxics Release Inventory (TRI) Public Data Release Report*, 'Executive Summary' and 'Chapter 5: TRI Data for Original Reporting Industries,' Released on April 11, 2001 along with the 1999 Toxics Release Inventory Data, (http://www.epa.gov/tri/tri99/pdr/index.htm).

US Federal Trade Commission, Statistical Report of the Bureau of Economics, *Annual Line of Business Report: 1974*, Washington, DC, September 1981.

US General Services Administration (1985, 1986, 1991, 1992) *Federal Register*. Washington, DC: US Government Printing Office.

US General Services Administration, *Federal Register*, vol. 59, no. 44, Monday, March 7, 1994 (Washington, DC: US Government Printing Office, 1994).

*Webster's Third New International Dictionary of the English Language Unabridged* (Springfield, Massachusetts: G&C Merriam Company, 1971).

# Appendices

## APPENDIX FOR CHAPTER 2:  THE EPA DATA

At the heart of Chapter 2 are data about air toxic emissions by industry at the four-digit Standard Industrial Classification (SIC) level.  The data were prepared under an EPA contract and made available to the public as the Crosswalk Air Toxic Emission data base through the US Department of Commerce's National Technical Information Service (US Environmental Protection Agency, 1991). The data are updated periodically, and the 1991 update used here made a special effort to focus on the list of hazardous air pollutants targeted in Title III of the 1990 Clean Air Act Amendment (Lynch, 1991, p. 2).  Further, these are the data that were available and would have conditioned the response of industry and government at the point in history on which this book focuses.

From the EPA data, I constructed data sets describing for each industry the chemicals that are associated with its production processes.  For this book, I have extracted just the information for manufacturing industries from the master data set that links industries and pollutants.   Of the total number (24,053) of observations of a pollutant associated with an industry category, about 87 percent (21,004 cases) were for manufacturing industries.

The *User's Manual* (Lynch, 1991) for the Crosswalk data base and the software (US EPA, 1991) focus on the emission factors (various measures of the amount of pollutant emitted into the atmosphere per unit of a polluting process) for the various chemicals.   The manual and the software are designed for users who typically would 'look up' information about a particular chemical.  Hence, until one 'dumps' the relevant files 'hidden' in the Crosswalk data base, the wealth of information linking chemicals to SIC industries is not at all transparent. That is because many chemicals do not have emissions factors in the data base, yet they are nonetheless in the data base with information about the SIC industries that use the chemicals in their production process.  The *User's Manual* says 'the data base currently contains . . . emission factors for over 300 compounds.'  Yet in fact several times that number of chemicals or groups of chemicals (roughly 1200 depending on how one counts the various families of chemicals) are catalogued if one's need is not for the emission factors and one looks at the entire data base.[1]

For two reasons, I do not want to focus exclusively on the cases with emissions factors in the EPA data. First, not all emissions factors are known. Second, even where that knowledge once developed shows that the toxic chemical is handled in a way or in a form that implies an emission factor of zero, the industry may well be the industrial location, in the chain of production from raw material to finished product, where research about any potential emissions problem is done. The potential problem might be in a vertically related industry, with the suppliers of an input or the purchasers of a product facing a potential emissions problem, yet the producers of the product may well provide R&D to solve the problem. For example, telephone conversations with the responsible executive at one major corporation revealed that the R&D to solve an emissions problem in the corporation's production process was being done by the supplier of the material that was causing the problem.[2]

Thus, a broad indication of the presence of a toxic chemical in the production process is desirable. Given the knowledge in the early 1990s, the details in the EPA data — conditioning the industrial R&D response at that point in industrial history — about emission factors are too spotty to provide the desired broad indication of potential problems with toxic chemicals.[3] Instead, the association of a toxic chemical with an industry's production processes is used for that broad measure. As it turns out, we shall see in subsequent chapters that the broader measure of the presence of toxic chemicals is positively and significantly correlated with the decision to do R&D to reduce Title III emissions in production.

Toxic air pollutants included in the data set as listed in 'Title III — Hazardous Air Pollutants' of Public Law 101-549 — Nov. 15, 1990 (Clean Air Act Amendments) follow. Note that Title III listed some catch-all categories (e.g., Chromium compounds or Lead compounds). For those catch-all categories, in the list below a separate heading is given followed by the set of chemicals or groups of chemicals studied under that heading. Brackets ([chemical name

---

[1] Since the *User's Manual* provides the internal file structure of every data file in the Crosswalk data base, one can efficiently pull out the large data sets linking chemicals that are toxic air emissions to the SIC industries. For many chemicals, a standardized chemical abstract (CAS) number is not available, and one must work with just the chemicals' names and their synonyms. Further, one must check for synonyms more generally — chemicals with the same CAS numbers although different names. Thus, creating the basic data set of chemicals and the industries that emit them required some work assembling and 'cleaning' the data.

[2] Confidentiality requirements prevent me from being more descriptive and stating exactly what the process was.

[3] As US EPA (2001) explains, the knowledge about emissions factors has continued to grow, with new chemicals being added to the group traced in the *Toxics Release Inventory* data.

CAS#]) denote chemicals for which no SIC manufacturing industry data were available (i.e., the chemical was not in the data set, or it was but there was no SIC data available for it, or it was but there were only nonmanufacturing SICs listed).

| Chemical Name | CAS Number |
|---|---|
| Acetaldehyde | 75070 |
| Acetamide | 60355 |
| Acetonitrile | 75058 |
| Acetophenone | 98862 |
| 2-Acetylaminofluorene | 53963 |
| Acrolein | 107028 |
| Acrylamide | 79061 |
| Acrylic acid | 79107 |
| Acrylonitrile | 107131 |
| Allyl chloride | 107051 |
| 4-Aminobiphenyl | 92671 |
| 3-Amino-2,5-dichlorobenzoic acid  (Chloramben) | 133904 |
| Aniline | 62533 |
| 2-Anisidine  (o-Anisidine) | 90040 |
| | |
| Antimony Compounds | |
| Antimony compounds, n.e.c. | |
| Antimony oxide | 1327339 |
| Antimony oxide ($Sb_2O_3$) | 1309644 |
| Antimony (V) pentafluoride | 7783702 |
| | |
| Arsenic Compounds (including Arsine) | |
| Arsenic compounds, n.e.c. | |
| Arsenic pentoxide | 1303282 |
| Arsenic trioxide | 1327533 |
| Arsine | 7784421 |
| | |
| Asbestos | 1332214 |
| Aziridine  (Ethylene imine) | 151564 |
| Baygon  (Propoxur) | 114261 |
| Benzene | 71432 |
| Benzidine | 92875 |
| Benzotrichloride | 98077 |
| Benzyl chloride | 100447 |
| | |
| Beryllium Compounds | |

| Chemical Name | CAS Number |
|---|---|
| Beryllium compounds | |
| | |
| Biphenyl | 92524 |
| Bis(2-chloroethyl) ether  (Dichloroethyl ether) | 111444 |
| Bis(chloromethyl) ether | 542881 |
| [Bromoethylene  (Vinyl bromide) | 593602] |
| Bromoform | 75252 |
| Bromomethane  (Methyl bromide) | 74839 |
| 1,3-Butadiene | 106990 |
| | |
| Cadmium Compounds | |
| Cadmium chloride | 10108642 |
| Cadmium compounds, n.e.c. | |
| Cadmium oxide | 1306190 |
| Octadecanoic acid, cadmium salt | 2223930 |
| | |
| [Calcium cyanamide  (Cyanamide, calcium salt (1:1)) | 156627] |
| Caprolactam | 105602 |
| Captan | 133062 |
| Carbaryl | 63252 |
| Carbon disulfide | 75150 |
| Carbon tetrachloride | 56235 |
| Carbonyl sulfide | 463581 |
| Catechol | 120809 |
| Chlordane | 57749 |
| Chlorine | 7782505 |
| Chloroacetic acid | 79118 |
| 2-Chloroacetophenone | 532274 |
| Chlorobenzene | 108907 |
| [Chlorobenzilate | 510156] |
| Chloroethane  (Ethyl chloride) | 75003 |
| Chloroform | 67663 |
| Chloromethane  (Methyl chloride) | 74873 |
| Chloromethyl methyl ether | 107302 |
| Chloroprene | 126998 |
| | |
| Chromium Compounds | |
| Chromate | 13907454 |
| Chrome tanned cowhide trimmings | 68131986 |
| Chromic acid | 11115745 |

Chemical Name                                                  CAS Number

Chromic acid, dipotassium salt                                 7789006
Chromic acid, lead(2+) salt (1:1)                              7758976
Chromic acid, strontium salt (1:1)                             7789062
Chromic acid (VI)                                              7738945
Chromic acid, zinc salt                                        13530659
Chromium compounds, n.e.c.
Chromium (III)                                                 16065831
Chromium (VI)                                                  18540299
Chromium (III) nitrate                                         13548384
Chromium (VI) oxide (1:3)                                      1333820
Chromyl fluoride                                               7788967
Dichlorodioxochromium                                          14977618
Dichromic acid, dipotassium salt                              7778509
Dichromic acid, disodium salt                                  10588019

Cobalt compounds
Cobalt (2+) sulfide                                            1317426
Cobalt compounds, n.e.c.

Coke-oven emissions
2-Cresol  (o-Cresol)                                           95487
3-Cresol  (m-Cresol)                                           108394
4-Cresol  (p-Cresol)                                           106445
Cresols (mixed isomers)                                        1319773
Cumene                                                         98828
Curene  (4,4-Methylene bis(2-chloroaniline))                   101144

Cyanide Compounds
Hydrogen cyanide                                               74908
Potassium cyanide                                              151508
Sodium cyanide                                                 143339

[DDE                                                           3547044]
[Diazomethane                                                  334883]
Dibenzofuran                                                   132649
1,2-Dibromo-3-chloropropane                                    96128
1,2-Dibromoethane  (Ethylene dibromide)                        106934
Dibutyl phthalate                                              84742
1,4-Dichlorobenzene                                            106467
3,3'-Dichlorobenzidine                                         91941

| Chemical Name | CAS Number |
|---|---|
| 1,1-Dichloroethane  (Ethylidene dichloride) | 75343 |
| 1,2-Dichloroethane  (Ethylene dichloride) | 107062 |
| 2,4-Dichlorophenoxy acetic acid  (2,4-D) | 94757 |
| 1,2-Dichloropropane  (Propylene dichloride) | 78875 |
| 1,3-Dichloropropene | 542756 |
| Dichlorvos | 62737 |
| Diethanolamine | 111422 |
| Diethyl sulfate | 64675 |
| [2,2-Diethyl-o-(p-nitrophenyl) ester phosphorothioic acid (Parathion) | 56382] |
| [1,6-Diisocyanatohexane  (Hexamethylene-1,6-diisocyanate) | 822060] |
| [3,3-Dimethoxybenzidine | 119904] |
| Dimethyl phthalate | 131113 |
| Dimethyl sulfate | 77781 |
| 4-Dimethyl aminoazobenzene | 60117 |
| Dimethylaniline, N,N- | 121697 |
| [3,3'-Dimethyl benzidine | 119937] |
| [Dimethyl carbamoyl chloride | 79447] |
| Dimethyl formamide, N,N- | 68122 |
| 1,1-Dimethylhydrazine | 57147 |
| 4,6-Dinitro-o-cresol | 534521 |
| 2,4-Dinitrophenol | 51285 |
| 2,4-Dinitrotoluene | 121142 |
| Dioctyl phthalate  (Bis(2-ethylhexyl)phthalate (DEHP)) | 117817 |
| 1,4-Dioxane | 123911 |
| 1,2-Diphenyl hydrazine | 122667 |
| 4,4'-Diphenyl methane diisocyanate (Methylene diphenyl diisocyanate (MDI)) | 101688 |
| Epichlorohydrin | 106898 |
| Ethyl acrylate | 140885 |
| Ethylbenzene | 100414 |
| Ethylene glycol | 107211 |
| Ethylene oxide | 75218 |
| Ethyloxirane  (1,2-Epoxybutane) | 106887 |
| Formaldehyde | 50000 |
| Glycol ethers | |
| Heptachlor | 76448 |
| Hexachlorobenzene | 118741 |
| Hexachloro-1,3-butadiene | 87683 |
| 1,2,3,4,5,6-Hexachlorocyclohexane, gamma-isomer (Lindane) | 58899 |

| Chemical Name | CAS Number |
|---|---|
| 1,2,3,4,5,5-Hexachloro-1,3-cyclopentadiene | 77474 |
| Hexachloroethane | 67721 |
| [Hexamethylphosphoramide | 680319] |
| Hexane | 110543 |
| Hydrazine | 302012 |
| Hydrogen chloride  (Hydrochloric acid) | 7647010 |
| Hydrogen fluoride  (Hydrofluoric acid) | 7664393 |
| Hydrogen sulfide | 7783064 |
| Hydroquinone | 123319 |
| [2-Imidazolidinethione  (Ethylene thiourea) | 96457] |
| | |
| Lead Compounds | |
| Lead compounds, n.e.c. | |
| Lead dioxide | 1309600 |
| Bis(acetato-O)tetrahydroxytrilead | 1335326 |
| Tetraethyl lead | 78002 |
| Tetramethyl lead | 75741 |
| | |
| Maleic anhydride | 108316 |
| | |
| Manganese Compounds | |
| Manganese compounds | |
| | |
| Mercury Compounds | |
| Mercury chloride | 7487947 |
| | |
| Methanol | 67561 |
| Methoxychlor | 72435 |
| 2-Methoxy-2-methyl propane  (Methyl tert butyl ether) | 1634044 |
| Methyl ethyl ketone | 78933 |
| Methyl iodide | 74884 |
| Methyl isobutyl ketone | 108101 |
| Methyl methacrylate | 80626 |
| [2-Methylaziridine  (1,2-Propylenimine) | 75558] |
| Methylene chloride | 75092 |
| 4,4'-Methylenedianiline | 101779 |
| [Methylhydrazine | 60344] |
| Methylisocyanate | 624839 |
| Mineral fibers | |
| Naphthalene | 91203 |

| Chemical Name | CAS Number |
|---|---|
| Nickel Compounds | |
| Nickel carbonyl | 13463393 |
| Nickel compounds, n.e.c. | |
| Nickel(II) oxide (1:1) | 1313991 |
| | |
| Nitran  (Trifluralin) | 1582098 |
| Nitrobenzene | 98953 |
| 4-Nitro-1,1'-biphenyl | 92933 |
| 4-Nitrophenol | 100027 |
| 2-Nitropropane | 79469 |
| Nitrosodimethylamine, N- | 62759 |
| 1-Nitroso-1-methylurea | 684935 |
| Nitrosomorpholine, N- | 59892 |
| [Pentachloronitrobenzene | 82688] |
| Pentachlorophenol (PCP) | 87865 |
| Phenol | 108952 |
| 1,4-Phenylenediamine | 106503 |
| [Phenyloxirane  (Styrene oxide) | 96093] |
| Phosgene | 75445 |
| Phosphine | 7803512 |
| Phosphorus (yellow or white) | 7723140 |
| Phthalic anhydride | 85449 |
| Polychlorinated biphenyls (PCBs) | 1336363 |
| Polycyclic organic matter | |
| Propane sultone | 1120714 |
| 1,3-Propiolactone  (beta-Propiolactone) | 57578 |
| Propionaldehyde | 123386 |
| Propylene oxide | 75569 |
| Quinoline | 91225 |
| Quinone | 106514 |
| Radionuclides  (including radon) | |
| | |
| Selenium Compounds | |
| Hydrogen selenide | 7783075 |
| Selenious acid, disodium salt | 10102188 |
| Selenium compounds, n.e.c. | |
| | |
| Styrene | 100425 |
| 2,3,7,8-Tetrachlorodibenzo-p-dioxin | 1746016 |
| 1,1,2,2-Tetrachloroethane | 79345 |

| Chemical Name | CAS Number |
|---|---|
| Tetrachloroethylene | 127184 |
| Titanium chloride (Titanium tetrachloride, TiCl4) | 7550450 |
| Toluene | 108883 |
| Toluene-2,4-diamine | 95807 |
| Toluene-2,4-diisocyanate | 584849 |
| 2-Toluidine (o-Toluidine) | 95534 |
| [Toxaphene | 8001352] |
| 1,2,4-Trichlorobenzene | 120821 |
| 1,1,1-Trichloroethane (Methyl chloroform) | 71556 |
| 1,1,2-Trichloroethane | 79005 |
| Trichloroethylene | 79016 |
| [2,4,5-Trichlorophenol | 95954] |
| 2,4,6-Trichlorophenol | 88062 |
| Triethylamine | 121448 |
| 3,5,5-Trimethyl-2-cyclohexen-1-one (Isophorone) | 78591 |
| [2,2,4-Trimethylpentane | 540841] |
| Urethane (Ethyl carbamate) | 51796 |
| Vinyl acetate | 108054 |
| Vinyl chloride | 75014 |
| Vinylidene chloride | 75354 |
| 1,2-Xylene (o-Xylenes) | 95476 |
| 1,3-Xylene (m-Xylenes) | 108383 |
| 1,4-Xylene (p-Xylenes) | 106423 |
| Xylenes (mixed isomers) | 1330207 |

## APPENDIX FOR CHAPTER 3: THE ORIGINAL QUESTIONNAIRE WITH COVER LETTER

*Letterhead of the Nelson A. Rockefeller Center for the Social Sciences at Dartmouth College*

August 12, 1993

TO: Director of Research & Development
      Company Name
      Address

  The Nelson A. Rockefeller Center for the Social Sciences at Dartmouth

College is sponsoring a study that I am conducting on the research response to the emission of toxic air pollutants in manufacturing. We would be most grateful if you could respond to the enclosed questionnaire. We need information about your company's research efforts that are designed to avoid problems with toxic air pollutants. The study aims to survey the extent of such research efforts in U.S. industry and to make recommendations for public policy to support such efforts. **All information given will be treated with the strictest confidentiality. No responses of individual companies will be released in any way at any stage of this research. Even industry aggregates will be suppressed if there are fewer than four companies in the total.**

Along with the questionnaire, I have enclosed a list of the chemicals on which I am focusing; the chemicals are those listed as hazardous air pollutants in Title III of the Clean Air Act Amendments of 1990. The survey inquires about your company's research related to air pollutants more generally as well as about research related to the Title III pollutants.

OUR MAIN INTEREST IS THE EXTENT TO WHICH YOUR COMPANY IS DOING RESEARCH THAT MAY ULTIMATELY REDUCE TOXIC AIR EMISSIONS. If possible we would also like your response to several subsidiary questions including the type of work (generic work on the problem, new processes, or new products), the organizational form of the research, the funding, the prospects for an adequate return on investment for the research and how those prospects affect the R&D decisions in this area, the percentage of the company's R&D effort going for such work, and the chemicals that are of concern in such work.

If you are unable to respond to all of the questions, we would nonetheless appreciate receiving your responses to those that you find most relevant. Please mail the questionnaire with your responses to:

> Professor John T. Scott
> c/o The Nelson A. Rockefeller Center for the Social Sciences
> Dartmouth College
> Hanover, NH 03755-3514

In return for your help, I will send you a copy of my completed report and would be pleased to answer any questions that you may have about the results.

If you have questions about the questionnaire or the survey, please feel free to call (telephone numbers) or fax (fax number). Thank you very much in advance for your help.

Sincerely,


John T. Scott
Professor of Economics

Enclosures:  Questionnaire, List of Title III Hazardous Air Pollutants


August, 1993

TO:

FROM:          Professor John T. Scott
               Department of Economics
               Dartmouth College
               Hanover, NH 03755-3514


               Telephone: (603) 646-2941
               Fax:  (603) 646-2122

**QUESTIONNAIRE**

The Research Response to the Emission of Toxic Air Pollutants in Manufacturing

A Study Sponsored by
The Nelson A. Rockefeller Center for the Social Sciences at Dartmouth College

We would be grateful if you would respond to the enclosed questionnaire that will provide helpful information about your company's research efforts designed to avoid problems with toxic air pollutants.  Along with the questionnaire, is a list of the chemicals on which the study is focused; the chemicals are those listed as hazardous air pollutants in Title III of the Clean Air Act Amendments of 1990. The survey inquires about research related to Title III air pollutants and also related to air pollutants more generally.  **All information given will be treated with the strictest confidentiality. No responses of individual companies will be released in any way at any stage of this research.  Even industry aggregates will be suppressed if there are fewer than four companies in the total.**

There are 13 questions about research on toxic air emissions specifically, 13 questions about development of new processes lessening toxic air emissions, 13 questions about development of new products lessening toxic air emissions, and a final question requesting chemical identification. We estimate that the entire questionnaire should not take longer than about 15 to 20 minutes to complete.

## PLEASE CIRCLE THE APPROPRIATE ANSWER OR ANSWERS.

1. Is your company conducting **research on any toxic air emissions** (those on the enclosed list **or** other air toxics)?     YES     NO

**If you answered NO to question 1, please skip to question 14.**

2. Approximately what part of your company's research on toxic air emissions focuses on those chemicals listed as hazardous air pollutants in Title III of the Clean Air Act amendments of 1990 (please see the list of Title III pollutants attached)?

ZERO     < 10%     10–20%     20–40%     40–60%     60–80%     80–90%     > 90%     ALL

3. Is your company's research on toxic air emissions

      STUDY OF THE TOXICITY OF THE EMISSIONS

      STUDY OF THE PROCESSES CREATING THE EMISSIONS

      OTHER (please specify if possible)

4. Is your company's research on toxic air emissions

      COMPANY FINANCED

      PERFORMED UNDER CONTRACT FOR ANOTHER COMPANY

      GOVERNMENT FINANCED

5. Is your company's research on toxic air emissions

      PERFORMED IN A COOPERATIVE VENTURE WITH OTHER FIRMS

      PERFORMED IN A COOPERATIVE VENTURE WITH GOVERNMENT

      PERFORMED INDEPENDENTLY

6. How difficult will it be to realize a normal (customarily expected given the riskiness of the projects) return from your company's investments in research on toxic air emissions?

NOT DIFFICULT     SOMEWHAT DIFFICULT     DIFFICULT     VERY DIFFICULT

7. How important is the ability to realize normal returns for your company's decision to

do research on toxic air emissions?

NOT IMPORTANT  SOMEWHAT IMPORTANT  IMPORTANT  VERY IMPORTANT

8.  Compared to your company's other research projects, is research on toxic air emissions

LESS COSTLY        ABOUT AS COSTLY        MORE COSTLY

9.  Compared to your company's other research projects, is research on toxic air emissions

LESS RISKY        ABOUT AS RISKY        MORE RISKY

10.  Compared to your company's other research projects, for research on toxic emissions are efficiencies from the size of the research effort

LESS IMPORTANT        ABOUT AS IMPORTANT        MORE IMPORTANT

11.  Is your company's research on toxic air emissions a response to a specific government request for information?        YES        NO

12.  Approximately what part of your company's total research and development effort for environmental projects is for research on toxic air emissions?

< 10%    10–20%    20–40%    40–60%    60–80%    80–90%    > 90%    ALL

13.  Approximately what part of your company's total research and development effort (for all projects, not just environmental ones) is for research on toxic air emissions?

< 10%    10–20%    20–40%    40–60%    60–80%    80–90%    > 90%    ALL

14.  Is your company conducting **R&D to develop new processes lessening toxic air emissions** (those on the enclosed list or other air toxics)?        YES        NO

**If you answered NO to question 14, please skip to question 27.**

15.  Approximately what part of your company's R&D on new processes to lessen toxic air emissions is concerned with those chemicals listed as hazardous air pollutants in Title III of the Clean Air Act amendments of 1990 (please see the list of Title III pollutants attached)?

ZERO    < 10%    10-20%    20-40%    40-60%    60-80%    80-90%    > 90%    ALL

16.  Is your company's R&D on new processes to lessen toxic air emissions for

PROCESS TECHNOLOGY TO BE USED BY YOUR FIRM IN ITS PRODUCTION

PROCESS TECHNOLOGY TO BE USED BY OTHERS IN THEIR PRODUCTION

PROCESS TECHNOLOGY EMBODIED IN A PRODUCERS GOOD TO BE SOLD

OTHER (please specify if possible)

17. Is your company's R&D on new processes to lessen toxic air emissions

COMPANY FINANCED

PERFORMED UNDER CONTRACT FOR ANOTHER COMPANY

GOVERNMENT FINANCED

18. Is your company's R&D on new processes to lessen toxic air emissions

PERFORMED IN A COOPERATIVE VENTURE WITH OTHER FIRMS

PERFORMED IN A COOPERATIVE VENTURE WITH GOVERNMENT

PERFORMED INDEPENDENTLY

19. How difficult will it be to realize a normal (customarily expected given the riskiness of the projects) return from your company's investments in R&D on new processes to lessen toxic air emissions?

NOT DIFFICULT   SOMEWHAT DIFFICULT   DIFFICULT   VERY DIFFICULT

20. How important is the ability to realize normal returns for your company's decision to do R&D on new processes to lessen toxic air emissions?

NOT IMPORTANT   SOMEWHAT IMPORTANT   IMPORTANT   VERY IMPORTANT

21. Compared to your company's other R&D projects, is R&D on new processes to lessen toxic air emissions

LESS COSTLY       ABOUT AS COSTLY       MORE COSTLY

22. Compared to your company's other R&D projects, is R&D on new processes to lessen toxic air emissions

LESS RISKY      ABOUT AS RISKY      MORE RISKY

23. Compared to your company's other R&D projects, for R&D on new processes to lessen toxic air emissions are efficiencies from the size of the research effort

LESS IMPORTANT       ABOUT AS IMPORTANT       MORE IMPORTANT

24. Is your company's R&D on new processes to lessen toxic air emissions a response to a specific government regulation?       YES       NO

25. Approximately what part of your company's total research and development effort

for environmental projects is for R&D on new processes to lessen toxic air emissions?

< 10%    10–20%    20–40%    40–60%    60–80%    80–90%    > 90%    ALL

26.  Approximately what part of your company's total research and development effort (for all projects, not just environmental ones) is for R&D on new processes to lessen toxic air emissions?

< 10%    10–20%    20–40%    40–60%    60–80%    80–90%    > 90%    ALL

27.  Is your company conducting **R&D to develop new products lessening toxic air emissions** (those on the enclosed list or other air toxics)?        YES        NO

**If you answered NO to question 27, please skip to question 40.**

28.  Approximately what part of your company's R&D on new products to lessen toxic air emissions is concerned with those chemicals listed as hazardous air pollutants in Title III of the Clean Air Act Amendments of 1990 (please see the list of Title III pollutants attached)?

ZERO    < 10%    10–20%    20–40%    40–60%    60–80%    80–90%    > 90%    ALL

29.  Is your company's R&D on new products to lessen toxic air emissions for

PRODUCTS PRODUCED WITH CLEANER PROCESS TECHNOLOGY

PRODUCTS THAT WILL HAVE LOWER TOXIC AIR EMISSIONS WHEN USED

PROCESS TECHNOLOGY EMBODIED IN A PRODUCERS GOOD TO BE SOLD

OTHER (please specify if possible)

30.  Is your company's R&D on new products to lessen toxic air emissions

COMPANY FINANCED

PERFORMED UNDER CONTRACT FOR ANOTHER COMPANY

GOVERNMENT FINANCED

31.  Is your company's R&D on new products to lessen toxic air emissions

PERFORMED IN A COOPERATIVE VENTURE WITH OTHER FIRMS

PERFORMED IN A COOPERATIVE VENTURE WITH GOVERNMENT

PERFORMED INDEPENDENTLY

32.  How difficult will it be to realize a normal (customarily expected given the riskiness of the projects) return from your company's investments in R&D on new

products to lessen toxic air emissions?

NOT DIFFICULT   SOMEWHAT DIFFICULT   DIFFICULT   VERY DIFFICULT

33. How important is the ability to realize normal returns for your company's decision to do R&D on new products to lessen toxic air emissions?

NOT IMPORTANT  SOMEWHAT IMPORTANT  IMPORTANT  VERY IMPORTANT

34. Compared to your company's other R&D projects, is R&D on new products to lessen toxic air emissions

LESS COSTLY      ABOUT AS COSTLY      MORE COSTLY

35. Compared to your company's other R&D projects, is R&D on new products to lessen toxic air emissions

LESS RISKY     ABOUT AS RISKY     MORE RISKY

36. Compared to your company's other R&D projects, for R&D on new products to lessen toxic air emissions are efficiencies from the size of the research effort

LESS IMPORTANT     ABOUT AS IMPORTANT     MORE IMPORTANT

37. Is your company's R&D on new products to lessen toxic air emissions a response to a specific government regulation?    YES    NO

38. Approximately what part of your company's total research and development effort for environmental projects is for R&D on new products to lessen toxic air emissions?

< 10%   10–20%  20–40%  40–60%  60–80%  80–90%  > 90%  ALL

39. Approximately what part of your company's total research and development effort (for all projects, not just environmental ones) is for R&D on new products to lessen toxic air emissions?

< 10%   10–20%  20–40%  40–60%  60–80%  80–90%  > 90%  ALL

40. (OPTIONAL — IF QUESTION 40 IS TOO BURDENSOME, PLEASE JUST IGNORE IT AND RETURN THE QUESTIONNAIRE WITH YOUR OTHER RESPONSES.) Could you please circle any chemicals on the attached list of hazardous air pollutants in Title III of the Clean Air Act amendments of 1990 that have been of concern in any of the research and development work described in your responses above?

THANK YOU FOR TAKING THE TIME TO RESPOND TO OUR QUESTIONNAIRE. PLEASE RETURN IT TO:

          Professor John T. Scott
          c/o The Nelson A. Rockefeller Center for the Social Sciences
          Dartmouth College
          Hanover, NH 03755-3514

Appendix to Questionnaire: Toxic Air Pollutants
Listed in 'Title III — Hazardous Air Pollutants' of Public Law
101-549 — Nov. 15, 1990 (Clean Air Act amendments)

| Chemical Name | CAS Number |
|---|---|
| Acetaldehyde | 75070 |
| Acetamide | 60355 |
| Acetonitrile | 75058 |
| Acetophenone | 98862 |
| 2-Acetylaminofluorene | 53963 |
| Acrolein | 107028 |
| Acrylamide | 79061 |
| Acrylic acid | 79107 |
| Acrylonitrile | 107131 |
| Allyl chloride | 107051 |
| 4-Aminobiphenyl | 92671 |
| 3-Amino-2,5-dichlorobenzoic acid  (Chloramben) | 133904 |
| Aniline | 62533 |
| 2-Anisidine  (o-Anisidine) | 90040 |
| Antimony Compounds | |
| Arsenic Compounds (including Arsine) | |
| Asbestos | 1332214 |
| Aziridine  (Ethylene imine) | 151564 |
| Baygon  (Propoxur) | 114261 |
| Benzene | 71432 |
| Benzidine | 92875 |
| Benzotrichloride | 98077 |
| Benzyl chloride | 100447 |
| Beryllium compounds | |
| Biphenyl | 92524 |
| Bis(2-chloroethyl) ether  (Dichloroethyl ether) | 111444 |
| Bis(chloromethyl) ether | 542881 |
| Bromoethylene  (Vinyl bromide) | 593602 |
| Bromoform | 75252 |
| Bromomethane  (Methyl bromide) | 74839 |
| 1,3-Butadiene | 106990 |
| Cadmium Compounds | |
| Calcium cyanamide  (Cyanamide, calcium salt (1:1)) | 156627 |
| Caprolactam | 105602 |
| Captan | 133062 |

| Chemical Name | CAS Number |
|---|---|
| Carbaryl | 63252 |
| Carbon disulfide | 75150 |
| Carbon tetrachloride | 56235 |
| Carbonyl sulfide | 463581 |
| Catechol | 120809 |
| Chlordane | 57749 |
| Chlorine | 7782505 |
| Chloroacetic acid | 79118 |
| 2-Chloroacetophenone | 532274 |
| Chlorobenzene | 108907 |
| Chlorobenzilate | 510156 |
| Chloroethane (Ethyl chloride) | 75003 |
| Chloroform | 67663 |
| Chloromethane (Methyl chloride) | 74873 |
| Chloromethyl methyl ether | 107302 |
| Chloroprene | 126998 |
| Chromium Compounds | |
| Cobalt compounds | |
| Coke-oven emissions | |
| 2-Cresol (o-Cresol) | 95487 |
| 3-Cresol (m-Cresol) | 108394 |
| 4-Cresol (p-Cresol) | 106445 |
| Cresols/Cresylic acid (isomers and mixture) | 1319773 |
| Cumene | 98828 |
| Curene (4,4-Methylene bis(2-chloroaniline)) | 101144 |
| Cyanide Compounds | |
| DDE | 3547044 |
| Diazomethane | 334883 |
| Dibenzofuran | 132649 |
| 1,2-Dibromo-3-chloropropane | 96128 |
| 1,2-Dibromoethane (Ethylene dibromide) | 106934 |
| Dibutyl phthalate | 84742 |
| 1,4-Dichlorobenzene | 106467 |
| 3,3'-Dichlorobenzidine | 91941 |
| 1,1-Dichloroethane (Ethylidene dichloride) | 75343 |
| 1,2-Dichloroethane (Ethylene dichloride) | 107062 |
| 2,4-Dichlorophenoxy acetic acid (2,4-D) | 94757 |
| 1,2-Dichloropropane (Propylene dichloride) | 78875 |
| 1,3-Dichloropropene | 542756 |
| Dichlorvos | 62737 |

| Chemical Name | CAS Number |
|---|---|
| Diethanolamine | 111422 |
| Diethyl sulfate | 64675 |
| 2,2-Diethyl-o-(p-nitrophenyl) ester phosphorothioic acid (Parathion) | 56382 |
| 1,6-Diisocyanatohexane  (Hexamethylene-1,6-diisocyanate) | 822060 |
| 3,3-Dimethoxybenzidine | 119904 |
| Dimethyl phthalate | 131113 |
| Dimethyl sulfate | 77781 |
| 4-Dimethyl aminoazobenzene | 60117 |
| Dimethylaniline, N,N- | 121697 |
| 3,3'-Dimethyl benzidine | 119937 |
| Dimethyl carbamoyl chloride | 79447 |
| Dimethyl formamide, N,N- | 68122 |
| 1,1-Dimethylhydrazine | 57147 |
| 4,6-Dinitro-o-cresol | 534521 |
| 2,4-Dinitrophenol | 51285 |
| 2,4-Dinitrotoluene | 121142 |
| Dioctyl phthalate  (Bis(2-ethylhexyl)phthalate (DEHP)) | 117817 |
| 1,4-Dioxane (1,4-Diethyleneoxide) | 123911 |
| 1,2-Diphenyl hydrazine | 122667 |
| 4,4'-Diphenyl methane diisocyanate (Methylene diphenyl diisocyanate (MDI)) | 101688 |
| Epichlorohydrin  (1-Chloro-2,3-epoxypropane) | 106898 |
| Ethyl acrylate | 140885 |
| Ethylbenzene | 100414 |
| Ethylene glycol | 107211 |
| Ethylene oxide | 75218 |
| Ethyloxirane  (1,2-Epoxybutane) | 106887 |
| Formaldehyde | 50000 |
| Glycol ethers | |
| Heptachlor | 76448 |
| Hexachlorobenzene | 118741 |
| Hexachloro-1,3-butadiene | 87683 |
| 1,2,3,4,5,6-Hexachlorocyclohexane, gamma-isomer (Lindane) | 58899 |
| 1,2,3,4,5,5-Hexachloro-1,3-cyclopentadiene | 77474 |
| Hexachloroethane | 67721 |
| Hexamethylphosphoramide | 680319 |
| Hexane | 110543 |
| Hydrazine | 302012 |
| Hydrogen chloride  (Hydrochloric acid) | 7647010 |

| Chemical Name | CAS Number |
|---|---|
| Hydrogen fluoride  (Hydrofluoric acid) | 7664393 |
| Hydrogen sulfide | 7783064 |
| Hydroquinone | 123319 |
| 2-Imidazolidinethione  (Ethylene thiourea) | 96457 |
| Lead Compounds | |
| Maleic anhydride | 108316 |
| Manganese compounds | |
| Mercury Compounds | |
| Methanol | 67561 |
| Methoxychlor | 72435 |
| 2-Methoxy-2-methyl propane  (Methyl tert butyl ether) | 1634044 |
| Methyl ethyl ketone (2-Butanone) | 78933 |
| Methyl iodide (Iodomethane) | 74884 |
| Methyl isobutyl ketone (Hexone) | 108101 |
| Methyl methacrylate | 80626 |
| 2-Methylaziridine  (1,2-Propylenimine) | 75558 |
| Methylene chloride (Dichloromethane) | 75092 |
| 4,4'-Methylenedianiline | 101779 |
| Methylhydrazine | 60344 |
| Methylisocyanate | 624839 |
| Mineral fibers | |
| Naphthalene | 91203 |
| Nickel Compounds | |
| Nitran  (Trifluralin) | 1582098 |
| Nitrobenzene | 98953 |
| 4-Nitro-1,1'-biphenyl | 92933 |
| 4-Nitrophenol | 100027 |
| 2-Nitropropane | 79469 |
| Nitrosodimethylamine, N- | 62759 |
| 1-Nitroso-1-methylurea | 684935 |
| Nitrosomorpholine, N- | 59892 |
| Pentachloronitrobenzene (Quintobenzene) | 82688 |
| Pentachlorophenol (PCP) | 87865 |
| Phenol | 108952 |
| 1,4-Phenylenediamine | 106503 |
| Phenyloxirane  (Styrene oxide) | 96093 |
| Phosgene | 75445 |
| Phosphine | 7803512 |
| Phosphorus (yellow or white) | 7723140 |
| Phthalic anhydride | 85449 |

| Chemical Name | CAS Number |
|---|---|
| Polychlorinated biphenyls (PCBs; Aroclors) | 1336363 |
| Polycyclic organic matter | |
| Propane sultone | 1120714 |
| 1,3-Propiolactone (beta-Propiolactone) | 57578 |
| Propionaldehyde | 123386 |
| Propylene oxide | 75569 |
| Quinoline | 91225 |
| Quinone | 106514 |
| Radionuclides (including radon) | |
| Selenium Compounds | |
| Styrene | 100425 |
| 2,3,7,8-Tetrachlorodibenzo-p-dioxin | 1746016 |
| 1,1,2,2-Tetrachloroethane | 79345 |
| Tetrachloroethylene (Perchloroethylene) | 127184 |
| Titanium chloride (Titanium tetrachloride, TiCl4) | 7550450 |
| Toluene | 108883 |
| Toluene-2,4-diamine | 95807 |
| Toluene-2,4-diisocyanate | 584849 |
| 2-Toluidine (o-Toluidine) | 95534 |
| Toxaphene (chlorinated camphene) | 8001352 |
| 1,2,4-Trichlorobenzene | 120821 |
| 1,1,1-Trichloroethane (Methyl chloroform) | 71556 |
| 1,1,2-Trichloroethane | 79005 |
| Trichloroethylene | 79016 |
| 2,4,5-Trichlorophenol | 95954 |
| 2,4,6-Trichlorophenol | 88062 |
| Triethylamine | 121448 |
| 3,5,5-Trimethyl-2-cyclohexen-1-one (Isophorone) | 78591 |
| 2,2,4-Trimethylpentane | 540841 |
| Urethane (Ethyl carbamate) | 51796 |
| Vinyl acetate | 108054 |
| Vinyl chloride | 75014 |
| Vinylidene chloride (1,1-Dichloroethylene) | 75354 |
| 1,2-Xylene (o-Xylenes) | 95476 |
| 1,3-Xylene (m-Xylenes) | 108383 |
| 1,4-Xylene (p-Xylenes) | 106423 |
| Xylenes (isomers and mixture) | 1330207 |

## APPENDIX FOR CHAPTER 4: CHEMICALS OF CONCERN IN THE TITLE III RESEARCH OF THE RESPONDENTS

There were 68 respondents who reported that Title III chemicals were the focus of some of their environmental R&D. Of those 68 respondents, 51 answered the optional question #40 and listed the Title III chemicals that are of concern in their environmental R&D. For the toxic air pollutants listed in 'Title III — Hazardous Air Pollutants' of Public Law 101-549 — Nov. 15, 1990 (Clean Air Act Amendments), the count of the number of respondents indicating the particular chemical is equal to the number shown in the following list in the rightmost column after the = sign, or equals zero if the chemical is shown in the list with the original questionnaire (in the appendix to Chapter 3) but is not shown here.

| Chemical Name | CAS Number | Count |
|---|---|---|
| Acetaldehyde | 75070 | = 2 |
| Acrylamide | 79061 | = 1 |
| Acrylic acid | 79107 | = 1 |
| 3-Amino-2,5-dichlorobenzoic acid (Chloramben) | 133904 | = 1 |
| Aniline | 62533 | = 1 |
| Antimony Compounds | | = 3 |
| Arsenic Compounds (including Arsine) | | = 7 |
| Asbestos | 1332214 | = 3 |
| Benzene | 71432 | = 10 |
| Benzidine | 92875 | = 1 |
| Benzyl chloride | 100447 | = 1 |
| Beryllium compounds | | = 2 |
| Bis(chloromethyl) ether | 542881 | = 1 |
| Bromomethane (Methyl bromide) | 74839 | = 2 |
| 1,3-Butadiene | 106990 | = 1 |
| Cadmium Compounds | | = 8 |
| Calcium cyanamide (Cyanamide, calcium salt (1:1)) | 156627 | = 1 |
| Carbon disulfide | 75150 | = 5 |
| Carbon tetrachloride | 56235 | = 6 |
| Carbonyl sulfide | 463581 | = 2 |
| Catechol | 120809 | = 1 |
| Chlorine | 7782505 | = 8 |
| Chloroform | 67663 | = 4 |
| Chloromethane (Methyl chloride) | 74873 | = 2 |
| Chloromethyl methyl ether | 107302 | = 1 |
| Chromium Compounds | | = 13 |
| Cobalt compounds | | = 1 |
| Coke-oven emissions | | = 1 |
| 2-Cresol (o-Cresol) | 95487 | = 3 |
| 3-Cresol (m-Cresol) | 108394 | = 2 |

| Chemical Name | CAS Number | Count |
|---|---|---|
| 4-Cresol  (p-Cresol) | 106445 | = 3 |
| Cresols/Cresylic acid (isomers and mixture) | 1319773 | = 1 |
| Cumene | 98828 | = 1 |
| Cyanide Compounds | | = 9 |
| Dibenzofuran | 132649 | = 2 |
| Dibutyl phthalate | 84742 | = 1 |
| 1,4-Dichlorobenzene | 106467 | = 1 |
| 3,3'-Dichlorobenzidine | 91941 | = 1 |
| 1,1-Dichloroethane  (Ethylidene dichloride) | 75343 | = 1 |
| 1,2-Dichloropropane  (Propylene dichloride) | 78875 | = 1 |
| Diethanolamine | 111422 | = 1 |
| Dimethyl phthalate | 131113 | = 1 |
| 4,6-Dinitro-o-cresol | 534521 | = 1 |
| Dioctyl phthalate  (Bis(2-ethylhexyl)phthalate (DEHP) | 117817 | = 2 |
| 1,4-Dioxane (1,4-Diethyleneoxide) | 123911 | = 1 |
| 4,4'-Diphenyl methane diisocyanate | 101688 | = 3 |
| (Methylene diphenyl diisocyanate (MDI)) | | |
| Epichlorohydrin  (1-Chloro-2,3-epoxypropane) | 106898 | = 1 |
| Ethyl acrylate | 140885 | = 1 |
| Ethylbenzene | 100414 | = 3 |
| Ethylene glycol | 107211 | = 9 |
| Ethylene oxide | 75218 | = 4 |
| Formaldehyde | 50000 | = 15 |
| Glycol ethers | | = 14 |
| Hexane | 110543 | = 4 |
| Hydrazine | 302012 | = 2 |
| Hydrogen chloride  (Hydrochloric acid) | 7647010 | = 12 |
| Hydrogen fluoride  (Hydrofluoric acid) | 7664393 | = 13 |
| Hydrogen sulfide | 7783064 | = 5 |
| Lead Compounds | | = 15 |
| Manganese compounds | | = 1 |
| Mercury Compounds | | = 6 |
| Methanol | 67561 | = 9 |
| Methyl ethyl ketone (2-Butanone) | 78933 | = 17 |
| Methyl iodide (Iodomethane) | 74884 | = 2 |
| Methyl isobutyl ketone (Hexone) | 108101 | = 8 |
| Methylene chloride (Dichloromethane) | 75092 | = 12 |
| 4,4'-Methylenedianiline | 101779 | = 1 |
| Methylisocyanate | 624839 | = 2 |
| Naphthalene | 91203 | = 2 |

| Chemical Name | CAS Number | Count |
|---|---|---|
| Nickel Compounds | | = 5 |
| Nitrobenzene | 98953 | = 2 |
| 4-Nitro-1,1'-biphenyl | 92933 | = 1 |
| 4-Nitrophenol | 100027 | = 1 |
| 2-Nitropropane | 79469 | = 1 |
| Pentachlorophenol (PCP) | 87865 | = 2 |
| Phenol | 108952 | = 4 |
| Phosgene | 75445 | = 2 |
| Phosphine | 7803512 | = 6 |
| Phosphorus (yellow or white) | 7723140 | = 1 |
| Polychlorinated biphenyls (PCBs; Aroclors) | 1336363 | = 5 |
| Polycyclic organic matter | | = 2 |
| Quinoline | 91225 | = 1 |
| Radionuclides (including radon) | | = 2 |
| Styrene | 100425 | = 5 |
| 2,3,7,8-Tetrachlorodibenzo-p-dioxin | 1746016 | = 3 |
| 1,1,2,2-Tetrachloroethane | 79345 | = 3 |
| Tetrachloroethylene (Perchloroethylene) | 127184 | = 10 |
| Toluene | 108883 | = 19 |
| Toluene-2,4-diamine | 95807 | = 1 |
| Toluene-2,4-diisocyanate | 584849 | = 2 |
| 2-Toluidine (o-Toluidine) | 95534 | = 1 |
| 1,1,1-Trichloroethane (Methyl chloroform) | 71556 | = 23 |
| 1,1,2-Trichloroethane | 79005 | = 7 |
| Trichloroethylene | 79016 | = 9 |
| Triethylamine | 121448 | = 3 |
| Urethane (Ethyl carbamate) | 51796 | = 1 |
| Vinyl acetate | 108054 | = 3 |
| Vinyl chloride | 75014 | = 4 |
| 1,2-Xylene (o-Xylenes) | 95476 | = 7 |
| 1,3-Xylene (m-Xylenes) | 108383 | = 5 |
| 1,4-Xylene (p-Xylenes) | 106423 | = 5 |
| Xylenes (isomers and mixture) | 1330207 | = 18 |

## APPENDIX FOR CHAPTER 8: THE COVER LETTER AND SURVEY INSTRUMENT FOR ASSESSING THE CHANGE IN ENVIRONMENTAL R&D

Letterhead

Date:

Company:

Address:

I am writing to follow up on the helpful response that your company provided to me on a research project.  In 1993, your company responded to a survey about industrial R&D efforts to reduce environmentally harmful emissions.  The person within your company who responded to the survey was _____.

Two papers were published using the survey responses.  Those papers were: (1) 'Environmental Research Joint Ventures among Manufacturers,' *Review of Industrial Organization*, vol. 11, no. 5 (October 1996), pp. 655–79.  (2) 'Schumpeterian Competition and Environmental R&D,' *Managerial and Decision Economics*, vol. 18 (1997), pp. 455–69.

I am currently updating the survey results, and I would be most grateful if you could complete the enclosed one-page survey update and return it to me at:

Professor John T. Scott
Department of Economics
Dartmouth College
Hanover, NH 03755

Responses will be treated as confidential, and only statistical aggregates (such as in the two published papers) that do not reveal individual responses will be released.  If you would like to receive the publications listed above, or any subsequent report based on the enclosed follow-up survey, please let me know and I shall be happy to send them to you.  I can be reached by regular mail, e-mail, telephone or fax, and if you would like to contact me with questions or if you would like for me to contact you, I shall of course be happy to talk with you.

Many thanks for your help with the original project, and of course for any help you can give with the update as well.

Sincerely yours,

John T. Scott

Enclosure:  Single-page survey follow-up form

Survey Update.[4]    Company Name:.

The purpose of this follow-up survey is to develop understanding of whether or not there has been a basic change in the proportion of environmental R&D in total industrial R&D since the initial survey in 1993. To measure such change, please approximate how your company's current ratio of environmental to total R&D compares to its early to mid-1990s value.[5]    Thus, for your company is (total environmental R&D)/(total R&D) currently smaller, or about the same, or greater than it was in the early to mid-1990s? If possible, please provide an approximate magnitude. For example, if the ratio has fallen, is it 50 percent of its earlier value. Or, if the ratio has risen, is it, for example, 150 percent of its earlier value. A circle at the relevant point on the following figure could be used to indicate the approximate range. For example, if the ratio is now half what it was, please circle 50 to indicate 50 percent; or, if the ratio has doubled, please circle 200 to indicate 200 percent. If the ratio now relative to the ratio earlier exceeds 2.5 (i.e., 250 percent), please just write the approximate ratio separately from the figure.

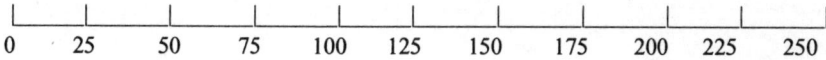

```
|____|____|____|____|____|____|____|____|____|____|
0    25   50   75   100  125  150  175  200  225  250
```

**Optional discussion**: If the type of R&D or the focus of your company's R&D has changed significantly since the early to mid-1990s, and if you have time to discuss the changes, please comment here. Also, if the environmental R&D that your company conducted in the early to mid-1990s led to patented processes or products, and if you have time to discuss any of them, please comment here or on a separate sheet.

[Space for response was provided]

**All information given will be treated with the strictest confidentiality. No responses of individual companies will be released in any way at any stage of**

---

[4] The survey is described in John T. Scott, 'Environmental Research Joint Ventures among Manufacturers,' *Review of Industrial Organization*, vol. 11, no. 5 (October 1996), pp. 655–79, and 'Schumpeterian Competition and Environmental R&D,' *Managerial and Decision Economics*, vol. 18 (1997), pp. 455–69.
[5] Research and development (R&D) expenditures are defined conventionally as in the surveys of industry by the National Science Foundation. By total R&D effort of your company, I intend company-financed and performed R&D, R&D performed under contract for others, and government-financed R&D performed by your company. Environmental R&D is defined as R&D to improve the environmental performance (reducing emissions, for example) of processes or products.

**this research. Even industry aggregates will be suppressed if there are fewer than four companies in the total.**

Many thanks for your help. Please return the completed form to Professor John T. Scott, Department of Economics, Dartmouth College, Hanover, NH 03755. Please indicate if you would like a copy of the publications. Thanks again. JTS

# Index